PRAISE FOR MILESTONES DELIVERABLES

Russ Finney, *Vice-President of Information Systems at Tokyo Electron America, Inc.* adds, "The entire company was astounded that such a complicated project could be so successful and delivered on time. The Pemeco implementation methodology, 'Milestone Deliverables', provided a comprehensible roadmap to follow with clearly-marked, tangible deliverables serving as signposts along the way... We moved to our new ERP systems on the final weekend and balanced our $51M of inventory to the penny!"

Vagesh Rajashekar, *Curtiss-Wright Corporation's Chief Information Officer*, comments, "With an incredibly complex product configuration process and complicated shop scheduling, Farris Engineering was concerned with being able to combine the technical challenges with the significant risks of a multi-site ERP implementation. *Pemeco's Milestone Deliverables* philosophy was followed as their senior consultants provided the project guidance, business understanding, and technical know-how. Through the many successful cut-overs and optimization phases, they vowed to "get down and dirty" with us, and they delivered on their promise! Pemeco's dedication ensured the success of our ERP implementation."

Fred Borah, *General Manager Peerless Instrument Company* states, "*Pemeco's Milestone Deliverables* approach enabled Peerless to meet targets in a timely fashion while documenting our processes effectively. I would recommend Pemeco's approach and team to any organization implementing management software."

Susan Alcide, *General Manager, Curtiss-Wright* "I never would have believed such a complicated project could be so successful. Pemeco's implementation methodology and emphasis on Milestone Deliverables was 'right on'!"

Ray Moran, *Dictaphone Sr. Director, Information Technology* "Large ERP implementations must be broken down into small, easy-to-manage segments. Pay careful attention to avoid scope-creep where unforeseen events distract the team's focus on planned deliverables. And hire an implementation partner ready to "get down and dirty" with the other team members. Pemeco provided these services to perfection."

Mike McNamara, *Automated Packaging Systems, Inc. Senior Vice President of Strategy and Business Analysis*, "Pemeco provided structure to the myriad of complex problems with a clear set of prioritized objectives, associated strategies, and a roadmap to get them done. Not only do their very senior consultants know business process, they are extremely adept communicators and managers."

Jim Leachman, *Enertech 's President*, comments, "With our hyper-aggressive schedule, project team members had to work efficiently, yet at a lightning-fast pace. Mr. Leachman continues, *Pemeco's Milestone Deliverables* methodology was a key to our success."

M. A. Chaudry, *Office of the County Executive, Kansas City, Missouri*, "The timing for evaluating Peter Gross' proven formula could not have been any better since we are at the early stages of developing our plans for ERP. We look forward to using the process shared by Peter Gross in managing the ERP project to achieve our goals and the Vision of the Jackson County Executive."

Jean-Guy Dugas, *General Manager of Goineau & Bousquet* adds, "Pemeco was the logical choice. Their skilled personnel understood our business, provided needed direction to our senior management, and managed the project in a top-notch fashion from training through implementation."

Richard Zieba, *Anachemia Science Ltd.'s president* comments, "One of the main reasons we chose Pemeco was the ability of its people to combine technical expertise with their years of business experience. The project progressed smoothly from beginning to end."

Ron Waxman, *President of Frisco Bay* adds, "Proper planning, controls, and organizational changes outlined in the plan and subsequently implemented were instrumental in helping us achieve these goals. We thank Pemeco for helping us to emerge from "the economic storm" a stronger and more profitable company."

Jean-Marie Bibeau, *President of Beau-Roc Inc.* adds, "Pemeco provided the project management skills and technical knowledge we required to complete the difficult transition."

Claude Lapointe, *Director of Operations, Herdt and Charton* comments, "The implementation took 12 months and, during this period, Herdt and Charton and Pemeco formed a cohesive team. We allocated the internal resources and Pemeco supplied the necessary technical and management skills needed to ensure the project's success."

Frank Engelberg, *President and owner of United Customs Brokers* comments, "Pemeco was our first choice to advise us in managing the project. Their personnel understand our business, guided our senior executives, and provided top-notch coordination and project support during the planning and implementation phases."

John Bertuccini, *Spartan Industries, Inc. Vice President Operations*, comments, "Pemeco was our first choice to manage the project. Their personnel understand both the technical as well as the human factors of computer system implementation. They learned our business and committed the right people to form a unified team with our employees."

Mitchell Herman, *Transco Plastics, Inc.* confirms, "Pemeco's hands-on involvement made all the difference to the success of the first phase of this project. And, of course, we now look forward to their assistance on phase two."

Thad Thomas, *IT Director, Jackson County, MO* . "Peter, thank you so very much for sharing your ERP implementation methodology with us. It certainly provided form and structure to a process that could otherwise be a risky endeavor."

Serge Le Guellec, *Chief Operating Officer at Mecaer Aviation International*, adds, "Pemeco's "Milestone Deliverables" philosophy was a key to our success. And, Pemeco's post "go live" support and dedication and flexibility during all project phases helped to ensure the success of our ERP implementation."

Louis Steen, *President of TEL EPION* adds, "Pemeco's 'Milestone Deliverables* was a key to our success. Pemeco's dedication ensured the success of our ERP implementation."

Trevor Allman, *the President and CEO of FusionIS*– an award winning Platinum Oracle partner – relied on *Pemeco's Milestone Deliverables* implementation methodology in successfully delivering a complex project for Johannesburg Metrobus - "the stock cutover from the old system to Oracle was 100% accurate. No items were lost and the item costs were transferred with 100% accuracy."

MILESTONE DELIVERABLES

—2ND EDITION—

ERP PROJECT MANAGEMENT METHODOLOGY

A HANDS-ON APPROACH TO LEADING SUCCESSFUL ERP IMPLEMENTATION PROJECTS

PETER GROSS, B.SC., M.ENG.

LIBRARY AND ARCHIVES CANADA CATALOGUING IN PUBLICATION

Gross, Peter
Milestone Deliverables ERP Project Management Methodology: A Hands-On Approach to Leading Successful ERP Implementation Projects

Second Edition

For permission requests and/or special sales, please contacting the publisher, Pemeco Consulting by phone at (647) 499-8161, by email at business@pemeco.com, or by visiting www.pemeco.com/milestone-deliverables-erp-implementation-project-management-book

Pemeco Consulting is a trade name of Pemeco Financial Holdings, Inc.

Published in Canada by Pemeco Financial Holdings, Inc.

Cover and illustrations by Stewart A. Williams Design, www.stewartwilliamsdesign.com

ISBN 978-0-9735893-1-3

TABLE OF CONTENTS

FOREWORD

The first edition of *Milestone Deliverables: The Hands-On Approach to Implementing Successful ERP Projects* was published in 2004. Since, operations management and enterprise software technologies have continued their fast-paced – and interrelated – innovation cycles.

Current operational and technology innovations

From a supply-chain perspective, omni-channel demands are pressuring companies to tighten their value chains, integrate their customer service channels, and interface with their business partners. And, modern technologies are providing organizations with the platforms to do so.

At the risk of dating this forward within a few years of publication, the following summarizes a few of the key technological innovations that are shaping today's enterprise software systems:

1. **Cloud computing**: arguably, this is the key enabler of many of today's current technology-related innovations. Cloud-based technologies offer users on-demand opportunities to access nearly limitlessly scalable computing resources.

2. **In-memory computing, business intelligence (BI), and big data analytics**: Together with cloud computing, these capabilities support rapid analysis of large sets of structured and unstructured data. Users are increasingly relying on analytics to improve predictive capabilities, and reduce uncertainties and associated volatility. From an innovation perspective, these types of technologies are triggering significant investment in artificial intelligence projects and academic disciplines relating to the data sciences.

3. **Mobility**: Mobility enhancements are increasingly responding to – and shaping – the realities of work-life, where job tasks are oftentimes performed away from one's desk. By offering real-time (or near real-time) transactional and analytical capabilities, mobile technologies are presenting companies with opportunities to become more agile and responsive.

4. **Internet-of-things (IoT)**: Increasingly, machines, devices, and equipment are being connected to central software command centers. Integrating "things" with planning and control software provides opportunities to improve automation, data accuracy, productivity, efficiency, and quality.

5. **Enterprise collaboration (a.k.a. the social enterprise)**: In an effort to institutionalize correspondence and improve productivity, organizations are turning to tools that facilitate internal and external communications, and contextual tracking of those communications for record-keeping purposes.

6. **User experience**: Design thinking – historically associated with consumer applications – is now a standard in the design of enterprise-grade applications. Elements such as responsive design, enhanced search capabilities, personalizeable interfaces, contextual help, and embedded learning management functionality are contributing to higher end-user adoption rates and lower onboarding costs.

ERP implementation project management: the more things change...

Notwithstanding these technology shifts, the recipe for delivering a successful ERP implementation has not changed.

For example, as the enterprise software market shifts from on-premise to cloud-based provisioning models, vendors and their service providers are increasingly pushing for remote, do-it-yourself implementation services. It is important for organizations to understand that the provisioning model doesn't change the fundamental nature of an ERP implementation – namely that it is a complex transformation project implicating organizational structures, people, processes, and technology. Successful delivery requires hands-on leadership, subject-matter expertise, and support.

Our firm's methodology – *Milestone Deliverables* – has been in practice for nearly 40-years. Over this period, we've used it (and seen it used) to implement card-index systems, mainframe systems, green-screen systems, GUI-systems, client-server systems, and, now, cloud-based systems. It's been used as a framework to implement Tier-I ERP systems at multi-national Fortune-500 enterprises, niche Tier-III ERP systems at small single-site companies, and organizations that fall somewhere in-between. Milestone Deliverables is also used to implement other enterprise software solutions, including configure-price-quote (CPQ), customer relationship management (CRM), quality management (QMS), warehouse management (WMS), and product lifecycle management (PLM) systems, among others. The testimonials reproduced in this edition are the evidence.

Milestone Deliverables is widely used across industries and solutions because it works. It breaks down big, hairy, audacious projects[1] into smaller, actionable, and measurable deliverables. And, each key deliverable acts as a signpost – or milestone – that indicates whether a project is on-course. With these, the team is always in a position to look at the signposts to orient (or, in some cases, re-orient) itself to the path to success. As is eloquently stated in the Introduction:

> *Pemeco's* Milestone Deliverables *premise is simple and powerful: If you provide managers and their teams with the ability to measure the project outputs, you will be rewarded with an organizational culture that is focused on deliverables.*

The methodology blends the following: project management, organizational change management, training, risk management, business process reengineering, system testing, and data migration. The disciplines are coherently woven into a project methodology that is comprised of 13 discrete phases with 14 measurable and actionable milestone deliverables that act as key indicators of success.

It took hundreds upon hundreds of projects to develop, test, and refine *Milestone Deliverables* to the point where the methodology appears to be simple. To this day, we continue to improve it. For example, since the book's first edition, Peter has refined the ERP Readiness Assessment and the Cutover Schedule. And, for our own client projects, we're now leveraging modern enterprise collaboration tools to facilitate project management, team communication, project health monitoring, and reporting.

Although we improve around the periphery, the core of **Milestone Deliverables** remains the same. The methodology strikes the right balance between agile and waterfall project management methods, business and technology, all bound by a wrapper of project control and organizational change management.

Extending *Milestone Deliverables* to our other practice areas

When I joined Pemeco in 2008, I was charged with building and formalizing complementary consulting practice areas. In doing so, it was important to embed in these new business lines the core tenets of *Milestone Deliverables*, namely: developing deliverables-based project methodologies with measurable outcomes that act as indicators of project success.

I'm proud to say that we have achieved this fundamental goal with respect to our business requirements assessment, enterprise software (including ERP) vendor evaluation and selection, and IT contract negotiation practice areas.

Our business requirements assessment methodology provides for systematic discovery and analysis of a company's organizational structure, business process, and technology needs that collectively feed into a project roadmap, schedule, and budget.

Our enterprise software evaluation and selection methodology provide a cohesive structure for the systematic evaluation of one or more vendors against defined business, IT, and market requirements relating to functionality, technology, usability, cost, and value.

Our IT contract negotiations methodology provides for a structured approach to negotiate deals that reflect the long-term nature of technology-related business partnerships.

To learn more about our services and capabilities, please visit www.pemeco.com, send us an email to business@pemeco.com, or call us at +1.647.499.8161.

In the meantime, enjoy the book and drive your project to success!

—*Jonathan Gross, LL.B., M.B.A.*

---

1. A shout-out to Jim Collins and Jerry Porras, whose BHAG – big hairy audacious goal – is the basis for this expression.

CHAPTER 1

INTRODUCTION

Enterprise Resource Planning (ERP) business software runs every aspect of a company including the management of orders, inventory, production, accounting, logistics, and people.

Project management of ERP implementations is an art, not a science. Although there is no clear-cut recipe or formula for success, statistics show that *over half of ERP implementation projects are perceived as failures.*

Since the late 1970's, scores of our firm's clients have implemented complex systems. As the projects got progressively more complex, we found we avoided floundering all over the place when we focused on the scope definition, rigorous change control, and a tightly monitored phase execution. To ensure that the project roadmap was clearly defined, we erected appropriate signposts along the way. From this analogy, the "Milestone Deliverables" methodology evolved, providing a framework that structures the myriad tasks into a simple, deliverable-oriented model.

Pemeco's Milestone Deliverables premise is simple and powerful: If you provide managers and their teams with the ability to measure the project outputs, you will be rewarded with an organizational culture that is focused on deliverables.

Each team member benefits from managing by deliverables. Working with their personalized work package in a simple intuitive framework, they gain instant clarity on the scope of their assignments and associated deliverables.

And the deliverables themselves are like a good wine – they constantly improve with age throughout the project cycle. As the team gathers more and more information and resolves any outstanding problems and issues, the deliverables evolve into better and more complete versions of themselves.

If truth be told, this "Milestone Deliverables" methodology is nothing more than a common-sense approach to managing people, objectives, and tasks. It has evolved continuously over 35 years and simply assumes that people are more effective, and better motivated, when working towards smaller, finite goals. Completion of these goals is signified by the production of tangible "end products." And with this feedback, managers are empowered to keep the project on track.

Faithful to its title, this book emphasizes the execution phases of an ERP implementation project as well as on the production of tangible milestone deliverables.

THE EMOTIONAL CURVE

People dynamics, important technical challenges, and the changing business climate are just a few of the variables that make each ERP project unique. However, our experience shows there is an underlying commonality called the project's "Emotional Curve." The project's managers and users must be made aware of this phenomenon early and often, as it affects the group's outlook and attitude during the entire project's life-cycle.

It's human nature to resist change. This tendency, plus the emotional highs and lows experienced during the project's phases, indicate that the emotional roller coaster can't be avoided. Fortunately, it can be anticipated, and this knowledge must be used to sensitize the entire organization at the outset. During the peaks and valleys, these feelings must be leveled out onto an even emotional keel to keep the team focused and productive.

At every opportunity, the team must be made aware of their current psychological state, as well as the upcoming ones. As an example, remind the team, "Things are working pretty well and we're all very excited now, but in a few weeks, we'll be sliding down a slippery slope. And we'll wonder why we ever left the security of our legacy systems." As reflected in Figure 1, the team's emotional dynamics are:

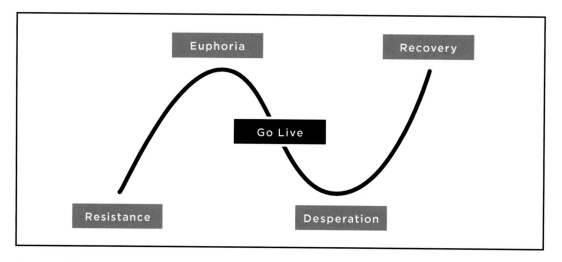

Figure 1. The "Emotional Curve"
The implementation team rides an emotional rollercoaster and the job of the project manager is to smooth out the highs and lows.

1. Resistance

People's fear of technology and their resistance to change are normal psychological re-actions in projects of any kind. ERP implementations touch every nook and cranny of a business, so it's natural to be afraid of the consequences.

During the Kickoff Presentation, one of the project manager's primary jobs is to un-derscore and empathize with this fear. The manager must reassure the team that this anxiety is only natural and that every team reacts this way. Our experience shows that only time will tell how successful the entire effort will be.

2. Euphoria

If the team has done its job well, euphoria will set in once the new business processes are presented at the completion of the Walkthrough phase. A marvelous epiphany oc-curs as the team clearly comprehends the benefits of the new system, the streamlined processes, and the improved reporting. They're now eager and thrilled at the prospect of using this new set of tools.

Again, we must remind them they are on the cusp of a frightening "descent into a dark place" as we identify many problems and issues that will force us to adapt our theo-ries to conform to a harsher reality. They will each have to increase their already burden-some work load as they test their theories through the piloting; they will need to commit enormous chunks of their time to the intense efforts of data migration and cutover.

3. Desperation

We collectively spiral downward into the depths as we work around all sorts of tribu-lations to shoehorn the new system into the business and vice-versa. Add to this the knowledge that we've pulled the plug on the trusted legacy systems! As we struggle to commit to memory how to use the new sessions in our day-to-day jobs, we wonder aloud, "Why'd we ever leave the security blanket of the past?"

This desperation and the understandable desire to return to a safe haven will last several months until the users become more at ease with new ways of doing their jobs. Again, we need to continually remind the team that this feeling of desperation is also temporary. As the "optimization" phase proceeds after cutover, things will improve, al-beit slowly, as we work our way up the curve towards a splendid "recovery."

4. Recovery

Although at first it is a supremely tough grind, new business processes become routine several months after turning on the new system. Senior management adapts to the new, improved tools to run the business, and the nagging, lower-priority issues unresolved after cutover are slowly, but surely, addressed.

THE RIGHT STUFF

This methodology focuses on the deliverables. But, for any ERP implementation to be a success, a foundation of solid fundamentals must be in place. Six of the most important are listed below:

1. All stakeholders must accept and spearhead the effort

ERP implementations are pervasive. They impact most departments and impose changes on the way people handle their day-to-day functions. If middle managers sense a lack of senior management support, they may introduce roadblocks that will adversely affect a project's performance.

Many senior executives regard ERP implementations as simple, albeit costly, technology upgrades. It is imperative they understand the end result will be a significant change to the way the organization looks and operates.

By the very nature of ERP systems, departments are forced to share information that they considered proprietary in the past. Stakeholders must insist that silos, constructed over the years for hoarding information, be dismantled.

Senior managers must facilitate negotiations among the various parties when disputes or disagreements erupt. They must keep their focus on the overall objectives and contribute sufficient time to the endeavor while avoiding being bogged down with the project's finer details.

2. Early input is required

Lack of user input will likely contribute to a bad ERP implementation. Introduce the project to those who will be affected by the outcome. Include not only the users, but also the business partners and other internal departments whose cooperation will be needed. Even though this may slow things down, the project management team must identify all the key resources needed to implement and support the ERP project.

As already mentioned, information must not be hoarded. Convince middle managers to be forthcoming about the way their departments run and alleviate their fears that the new software will reduce their spheres of influence.

Senior managers must reinforce the project's benefits and stress the importance of sharing information. They must make sure three broad groups contribute before the project gathers steam: those who will be affected by it, those who will implement it, and those who will pay for it.

3. Well-defined specifications and change control must be present

Poorly defined specifications and a lack of change control procedures are prime causes of ERP project failure. Requirements must be well defined up front to obtain the required consensus among the stakeholders.

One of the keys is to secure input from the stakeholders through a series of planning meetings to summarize in clear terms what the project can, and cannot do.

Senior management must ensure project scope changes are managed in a formal manner. This includes, but is not limited to, delays in the schedule or requests for additional money.

4. Set realistic expectations

Estimating ERP project schedules and resource requirements has always been a hit-and-miss affair. Stakeholders, less knowledgeable about what the technology can really do, create their own expectations — even fantasies.

If expectations are not set, scope creep is inevitable. An initially straightforward project can evolve into an unmanageable one, violating schedules and consuming resources.

A formal project charter must be established to set expectations. Project management must ensure that strict budgeting and risk assessment occur while senior management makes sure the culture is in place for a strong project management discipline.

Projects fail, not because the tasks are insurmountable, but because they're engendered by an effort to transform the company. Information Technology is used as the catalyst for that change and makes a very convenient scapegoat if things turn ugly. When a project falls short, it may look like IT failed—but it's almost always because the organizational change was unsuccessful.

4. Choose consulting partners wisely

Employees often experience resistance and resentment when outsiders are paid to do consulting work within a company. And this is an impediment to sharing information.

Avoid the bait-and-switch practice wherein the integrator's "A Team" flies in for the pre-sales meetings and launch — and then promptly hands the project off to a less experienced and, perhaps, less-skilled, crew. And when working on the consulting contract, it's important to negotiate toward a reasonable, achievable agreement. These aren't adversaries; these are partners.

5. Communicate, communicate, communicate

The everyday communication problem is worse when IT is involved, simply because it's hard for a lay person to grasp the lingo. Use non-technical terminology whenever possible, especially when communicating outside the project team.

The project manager must be forthcoming with any news good or bad. Line workers don't want to be the bearers of bad news, and senior managers contrive to not hear bad news if it's ever delivered. As a result, nobody sounds the alarm on projects that have "disaster" written all over them until it's too late.

Ensure that senior executives are available when they're needed and that they stay in constant touch with the project management.

CHAPTER 2

PROJECT TEAMS

Project managers must, and I repeat *must*, acquire the best people available for the implementation team; then they must do whatever it takes to keep the flotsam out of their way. By acquiring the best people — the most skilled, the most experienced, the best qualified — they can often compensate for too little time or money or other project constraints. A manager should serve as chief advocate for these valuable team members, helping to protect them from outside interruptions, and enabling them to acquire the tools they need to apply their talents.

In some cases it becomes necessary to lean on the heads of the business units to commit their strongest people. Make it clear that you understand the implications of doing so: a sharp, but temporary, drop in business performance.

Refer to Figure 2 for a view of a sample project organization.

THE CORE TEAM

Project teams are often made up of a project manager, a set of key users or functional leads, selected end users, and (usually) outside consultants who provide functional and technical support to the team. This group is referred to as the "Core Team."

The project managers and key users must possess the following qualities, in order of importance:

1. A general business knowledge and related operational experience
2. Strong leadership and delegation skills, plus willingness to roll up their sleeves and work down in the trenches
3. Excellent verbal and written skills, and the ability to communicate effectively from the level of senior management down to the shop floor
4. The consultants must possess significant product-related functional and technical knowledge

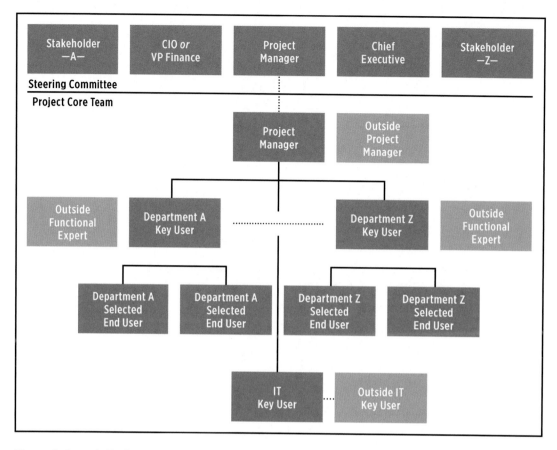

Figure 2. Sample Project Organization
Team members must be the best and brightest stars in the organization.

In an ideal world, core team members are assigned full-time to the ERP project and are back-filled in their day-to-day, "real" jobs. If this isn't possible, insist on a minimum of 50% of their time dedicated to the project. As project manager, you must be prepared to fight tooth and nail for this time as business needs will surely conflict with the project's needs.

FUNCTIONAL TEAMS

Most companies can be divided into functional, or departmental, units. For example, a department in a manufacturing concern might be broken down in this way:
- Engineering
- Marketing
- Sales and Customer Service
- Supply Chain

- Planning
- Purchasing
- Manufacturing
- Quality
- Warehousing
- Service Management
- Finance

These functional units can be further sub-divided depending on the complexity of the business processes within the unit. For example, Engineering may be sub-divided into Design and Manufacturing Engineering, and Service Management may be sub-divided into Call Registration, Dispatch, and Repair Center.

The data residing in the ERP system is often the responsibility of many departments; maintenance of this data must be coordinated by a separate functional unit, the Competency Center. Once the new system is in production, the members of this Competency Center must also drive the ongoing system and process improvements.

The changing business practices imposed by the new ERP application will affect the work scope and habits of most employees. In many cases, new skills will be needed, and the process of re-training can be a significant undertaking. This falls under the responsibility of a Change Management functional unit.

Each functional unit (and sub-division) must have its own implementation team comprised of:

- a core team key user assigned to direct the team
- a number of selected end users covering all the unit's business processes and possessing an adequate knowledge of other department operations
- a functional consultant with a generic understanding of the unit's processes and a detailed knowledge of the software itself

Depending on the team's size, key users may be responsible for more than one functional unit at a time.

RESPONSIBILITY AND AUTHORITY

Project managers must focus on three dimensions of a project's success. Simply put, project success means completing all project deliverables: 1) on time; 2) within budget; and 3) achieving a level of quality that is acceptable to sponsors and stakeholders. The project manager must keep the team's attention focused on achieving these broad goals.

Because projects are finite endeavors with limited time, money, and other resources, they must be kept moving resolutely toward completion. Most team members have other priorities, but it's up to the project manager to keep their attention focused on the project deliverables and deadlines. Although they are time-consuming, regular status meetings and reminders are essential.

Projects must have enough time to "do it right, the first time." And project managers must fight for this time by demonstrating to sponsors and senior management why it's necessary and how the time spent will result in quality deliverables.

The project manager's responsibilities must be matched by an equivalent authority. It's not enough to be held responsible for project outcomes; project managers must ask for, and obtain, enough authority to execute their responsibilities. Specifically, managers must have the authority to: acquire and coordinate resources; request and receive senior management's cooperation; and make appropriate, binding decisions which have a lasting impact on the success of the project.

Most project sponsors and stakeholders rightfully demand the authority to approve the project deliverables, either wholly or in part. Along with this authority comes the responsibility to be an active participant in the early stages of the project. They will help define the deliverables; complete timely reviews of the interim deliverables to keep the project moving; expedite the project manager's access to senior management and members of the target audience; and facilitate essential documentation.

THE STEERING COMMITTEE

The ERP Steering Committee is a key communication channel. It should be composed of the Chief Executive (or equivalent), the business-manager stakeholders at an EVP or VP level, the CIO, and the ERP Project Manager. The term business manager is used intentionally to represent those managers responsible for organizations that may benefit or be affected by any ERP system modules. *It must be remembered that ownership still resides with the respective business areas.* Each member of this committee must understand the company's strategic goals and budget constraints, and must have the authority to make decisions regarding resources and enterprise-level priorities.

This committee offers the highest-level strategic guidance, including a review of the ERP Plan from which detailed implementations will be developed. The ERP plan includes details of the resources required, their associated costs, and timelines. It identifies potential risks and challenges not previously considered; for this reason, the ERP Steering Committee should continually review and assess progress along the way.

Key to the project's success will be the commitment by senior management to ensure that the business is truly driving all enterprise-level technological implementations.

CHAPTER 3:

IMPLEMENTATION OF THE MILESTONE DELIVERABLES PROCESS

A Milestone Deliverables ERP implementation project is made up of a "lucky thirteen" basic phases. However, these phases can be expanded upon and further broken down into smaller and smaller components. In fact, that is exactly what is needed to link tangible deliveries to achievable tasks.

Phases are made up of tasks and deliverables, and some cannot start until their predecessors complete. The deliverables are strategically associated with key milestones; ultimately when all tasks and deliverables are completed, the project is technically over.

OVERVIEW OF THE PHASES

Like the pyramid in Figure 3, the phases build on a foundation of solid planning. They then move to the development of concepts; proving concepts; transferring the knowledge learned; and taking the data used in the old systems and transferring it to the new.

The logical order of the phases enables the deliverables to evolve naturally, and incrementally, as the team's understanding of the ERP application software improves.

Establish the Objectives and Guidelines (Planning Phase)
The first stage of any project is to define the scope, set expectations, prepare high-level plans, estimate costs, and develop timelines.

Develop the Concepts (Key User Training, Mapping, and Walkthrough Phases)
Then, learn the new systems, revise the business processes, and ensure (in theory) that they work seamlessly across the organization.

Prove the Concepts (Conference Room Pilot, Departmental Pilot, and Integrated Pilot Phases)
Pilot several layers of testing to prove that the theories can be put into practice on the new ERP software.

Transfer the Knowledge (End User Training Phase)
Enhance worker skills and train the end users to run the business on the new system.

Out with the Old (Migration, Customizations, Interfaces, Special Projects, IT Infrastructure Phases)
Bring the data from the old systems to the new, adapt and develop ancillary systems. Construct the appropriate IT infrastructure.

In with the New (Cutover Phase)
Close down the old, and fire up the new ERP system.

Improve Forever (Optimization Phase)
Never stop improving the ways of conducting business.

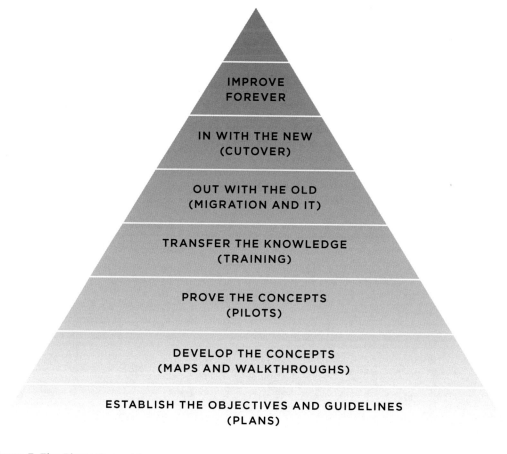

Figure 3. The Phase Pyramid
A solid foundation, built from the bottom up, ensures the project will not crumble under its own weight.

Figure 4 shows a view of the time-line of the relationship of these phases. In the sample, a six-month start-to-finish duration is used for reasons of brevity and simplicity.

Each of the thirteen phases, their tasks, and associated deliverables is covered later in this chapter.

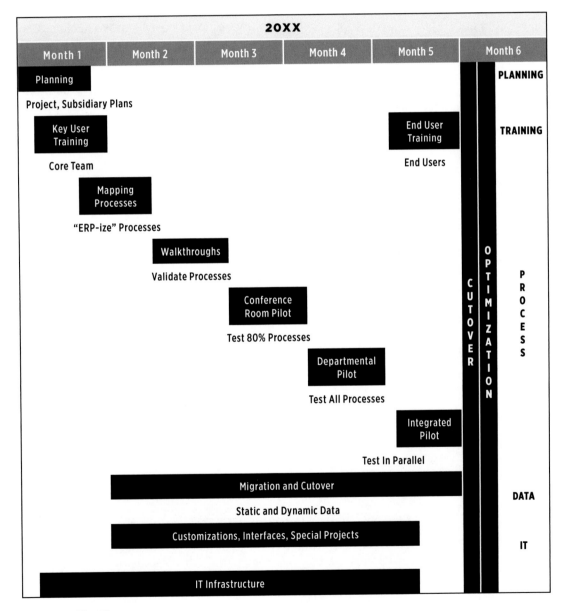

Figure 4. The Phases
Thirteen phases comprise a complex ERP implementation project.

OVERVIEW OF THE TASKS AND MILESTONE DELIVERABLES

The project tasks are accompanied by fourteen tangible deliverables, strategically sprinkled throughout the project's phases. Completing a deliverable signifies the completion of a particular milestone. And this completion often is a prerequisite to the start of the next phase.

Figure 5 and Figure 6 illustrate the further breakdown of the major phases described in the previous section. The high-level tasks and associated Milestone Deliverables completion points are also shown on the chart.

Each "bubble" (   ) contains a description of the deliverable and points to the milestone event after which the deliverable must be completed. Note that the same deliverable can appear several times since many deliverables repeat; deliverables are constantly improving while being revised over the life of the project.

The format of these deliverables is not set in stone. In fact, this methodology does not intend to limit the choice of software for word processing, spreadsheets, project management or databases, nor does it intend to suggest using the sample report formats without modification. What is important is to be faithful to the concept of using this methodology to manage the team to produce high-quality, tangible results in a timely fashion.

The Milestone Deliverables are:

1. **Project and Subsidiary Plans** to list project objectives and scope
2. **Kickoff Presentation** to signal the start of implementation execution
3. **Core Team Training Courses** to teach ERP overviews and details
4. **Business Scenario Lists** to list each department's processes
5. **Blueprint White Papers** to document each department's operations
6. **Gaps and Issues Database** to assist tracking and reporting gaps and issues
7. **Change Management Plan** to lay out end users' skills upgrades
8. **Walkthrough Presentations** to present each department's business processes
9. **80% and 20% Scenario Scripts** to detail user instructions for each business scenario
10. **User Training Courses** to teach the new operating methods and ERP application details
11. **User Documentation** to combine the many deliverables into a complete document
12. **IT Specifications** to detail customizations, interfaces, and special projects
13. **Migration Plan** to map out tasks for conversion and entry of legacy data
14. **Cutover Plan** to document the project final weeks' tasks

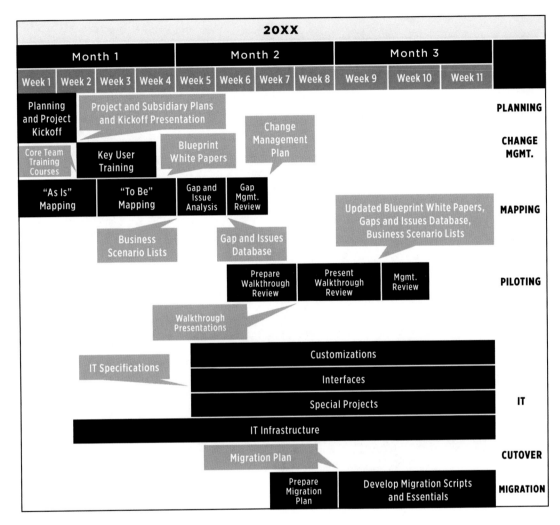

Figure 5. High-level Tasks and Milestones Deliverables

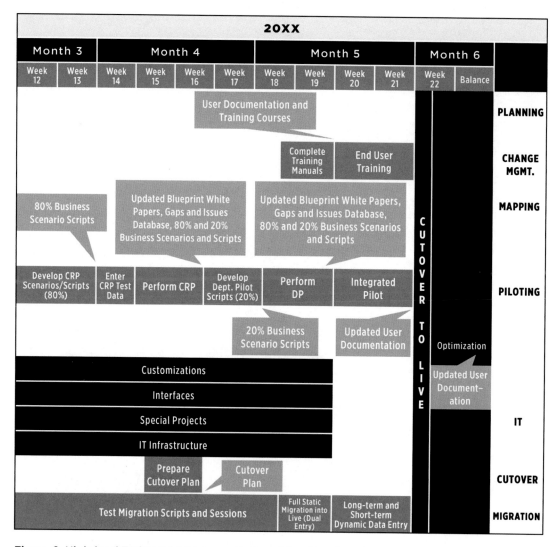

Figure 6. High-level Tasks and Milestones Deliverables (Continued)

THE PROJECT SCHEDULE OVERVIEW

In the samples, we have used an elapsed six-month project timetable for the sake of simplicity. In fact, full-scale ERP projects can vary in length from brief 4 ½-month implementations to larger undertakings lasting from one to several years.

Figure 7 shows the collapsed view of all the project phases. In this example, the first task starts in early June and the final Optimization Phase completes just less than six months later.

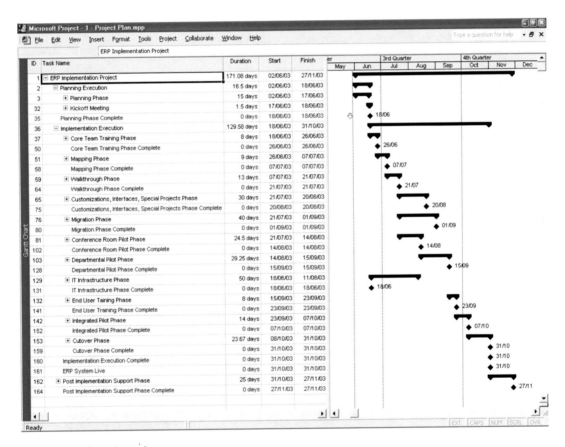

Figure 7. Project Overview

PHASES, TASKS AND
MILESTONE DELIVERABLES

The Planning Phase

Proper planning sets expectations and establishes a solid foundation for the entire project. Before rushing into the execution phases, make sure the assumptions, strategies, and high-level project roadmap are well documented.

The Timeline

Planning is the project's opening phase. Prepare comprehensive plan documents and memorable presentations.

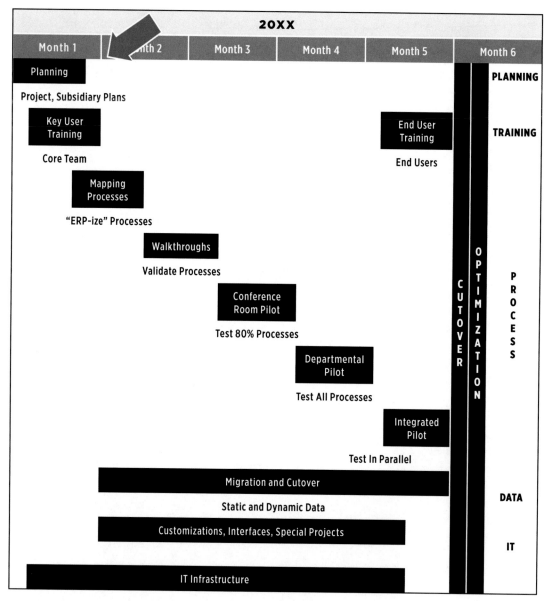

What this Phase Accomplishes

The project and subsidiary plans are developed during this phase. These are consolidated into a presentation and delivered to the steering committee for approval.

The project plan is a management document used to guide the project's execution and control. Like other documents, it is developed by continuously and repeatedly moving from generic requirements and undated events to specific detailed needs, explicit resource allocations, and precise dates.

The project plan is used to:

- Document major assumptions
- Provide a baseline for project measurement and control
- Guide the project's execution
- List alternative strategies and final recommendations
- Facilitate communication among stakeholders

The Kickoff Meeting signals the start of the project's execution phases and is usually given to the entire core team and company staff. Please refer to the Kickoff Meeting section in Chapter 5 for a detailed discussion.

The Tasks

1. Meet with the project stakeholders.
   Obtain the information necessary to prepare the project plan.
2. Prepare the project and subsidiary plans.
3. Prepare the steering committee presentation.
   Summarize the project and subsidiary plans for a formal presentation to the steering committee.
4. Present the plans to the steering committee.
5. Prepare and stage the project Kickoff Meeting.

   Summarize the project and subsidiary plans for a formal project Kickoff Meeting.

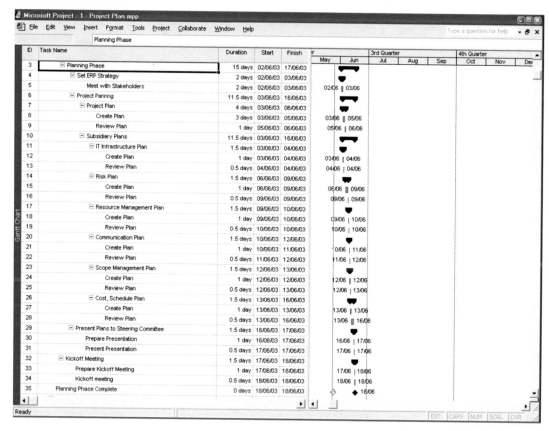

Figure 8. Planning Phase

The Deliverables

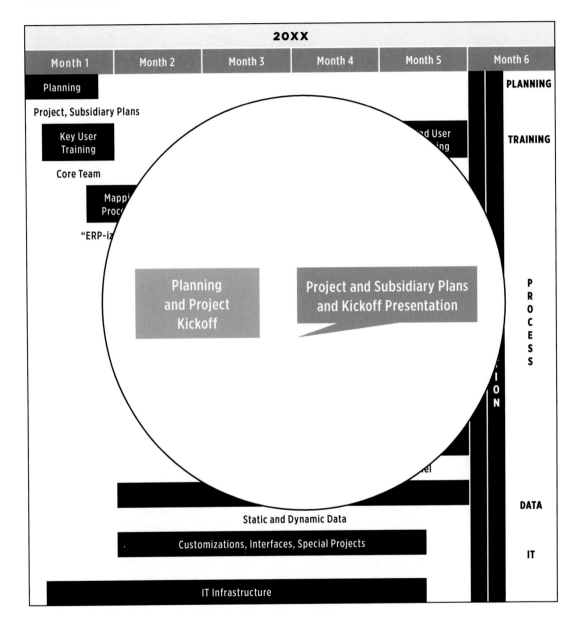

The First Deliverable: The Project and Subsidiary Plans

At a minimum, the following should be included in the Project Plan:

i)   The Project Charter: Document the business rationale for wanting to implement a new ERP system.

ii) The Scope Statement: Establish measurable success factors (MSF) and associated strategic business accomplishments (SBA) to deliver the end business results the stakeholders want.

iii) Provide a milestone level overview of the target dates and the project costs.

iv) Identify the project organization and the key staff requirements.

v) Include subsidiary plans or chapters such as:

(1) The IT Infrastructure and Procurement Plan

Present the ERP system's impact on:

   b) Hardware and system software
   c) Communications hardware and software
   d) Networks
   e) The disaster recovery plan
   f) Staffing levels and workload
   g) Define a sub-project team structure
   h) Identify phases and associated tasks, resources, and timelines
   i) Identify a procurement list of capital and recurring cost items

(2) The Risk Plan

Brainstorm to identify potential problems, ones that aren't guaranteed to occur. Areas to consider are: possible weak areas such as unknown technology; critical variables, such as the timely delivery of a vendor's database software; problems that have plagued past projects, such as loss of key staff, missed deadlines or error-prone software.

Analyze each risk item and render it very specific. Set priorities and determine where to focus risk mitigation efforts. Some of the identified risks are unlikely to occur, and others may not be serious enough to warrant concern. During the analysis, discuss each risk item to determine its potential damage and likelihood of occurrence.

(3) The Cost and Schedule Plan

Present the internal and external costs, and projected timelines detailed to the task level.

(4) The Resource Management Plan

Identify, document, and assign project roles, responsibilities, and reporting relationships. Describe when and how human resources will be back-filled, brought in and taken off the project teams. Detail how the individual and group competencies will be developed (training, etc.). Where the corporate culture permits, establish a reward system tied to project performance (perhaps using some of the deliverables). Attempt to co-locate the core team members while working on project activities.

(5) The Communication Plan

List the methods of collecting and disseminating updates and corrections to previously distributed material. Establish a distribution structure that details to whom minutes, status reports, etc., will flow; and provide a description of the material's format, content, and conventions. Include both the internal and external target audience.

(1) The Scope Management Plan

Describe how scope changes will be identified, classified, and integrated into the project.

vi) Identify open issues and pending decisions.

The Second Deliverable: The Kickoff Presentation

In Chapter 5, we will discuss the kickoff meeting and its presentation.

The Core Team Training Phase

The essential first step of the project construction is to effectively train the core team. Training is the foundation upon which the core team builds the new business processes from existing ones. The application software vendor or the consultants deliver the courses.

The Timeline

The mapping process that follows depends heavily on the quality of the training delivered during this phase.

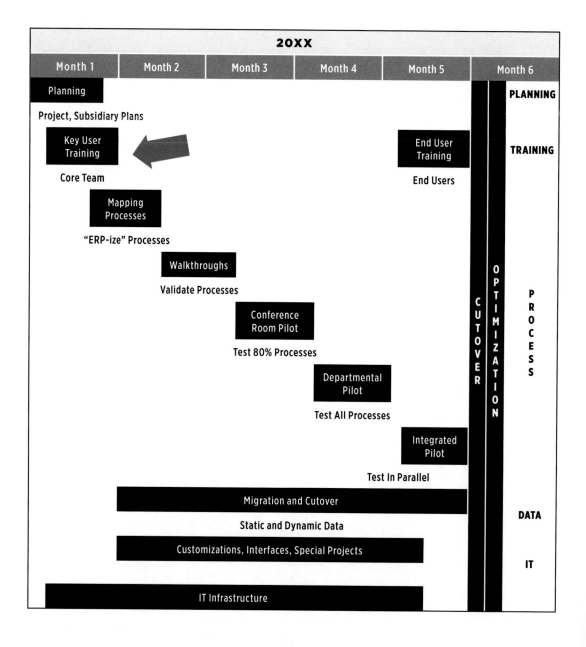

What this Phase Accomplishes

By the end of this phase, the core team members will have mastered the fundamentals of ERP theory as well as specific details of the new ERP application.

This is accomplished by providing the following training sessions to the core team:

ERP fundamentals and ERP application overviews
Hands-on navigation
Application-specific detailed functional training
ERP application administrative training

The Tasks

(1) Prepare the training sessions

Our experience shows that targeted training is more beneficial than generic sessions. For that reason, consultants are often better suited to tailor the training to the specific business environment.

(2) Deliver the following sessions:

c) ERP fundamentals and an application-specific overview training

Fundamentals training, adapted to the core team's knowledge level, discusses general ERP theories. This is followed by application-specific instruction covering the software modules and their interrelationships. The audience: the entire core team.

d) Hands-on navigation training

Navigation training provides specific instruction on the manipulation of fields and the navigation of screens, forms, and sessions. The audience: the entire core team.

e) Application-specific detailed functional training

These training courses teach the functional details of the ERP application modules. The audience: the core team members of the appropriate functional unit.

f) ERP application administrative training

This training teaches concepts relative to the administration of the application software. Administrative function should include, but are not limited to, application of software patches, revisions, and releases; maintenance of all levels of access security; printers and screen handling, etc. The audience: the IT core team members.

After the training, the core team must be sufficiently knowledgeable on the new ERP application to begin the Mapping Phase where they will adapt the existing business flows to the new ERP environment.

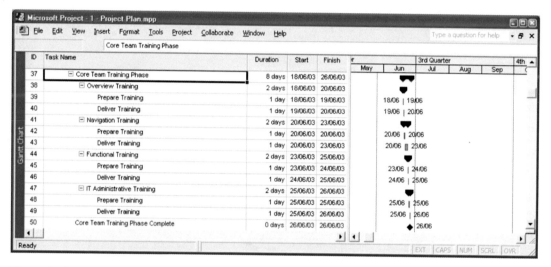

Figure 9. Core Team Training Tasks

The Deliverables

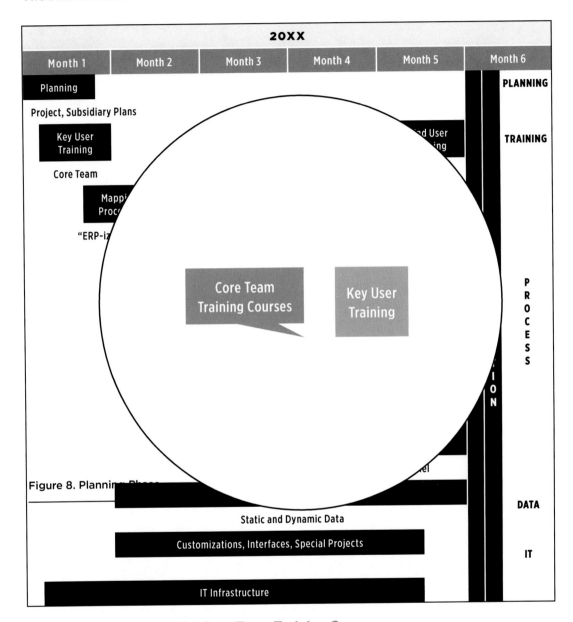

Figure 8. Planning Phase

The Third Deliverable: The Core Team Training Courses

Many ERP Overview Training sessions are available. We have provided a partial sample to illustrate the use of high-level diagrams to guide the delivery of the course content. As the remaining courses are specifically tailored to the project, business, and ERP application environment, there are no sample deliverables provided in this text.

ERP CYCLE

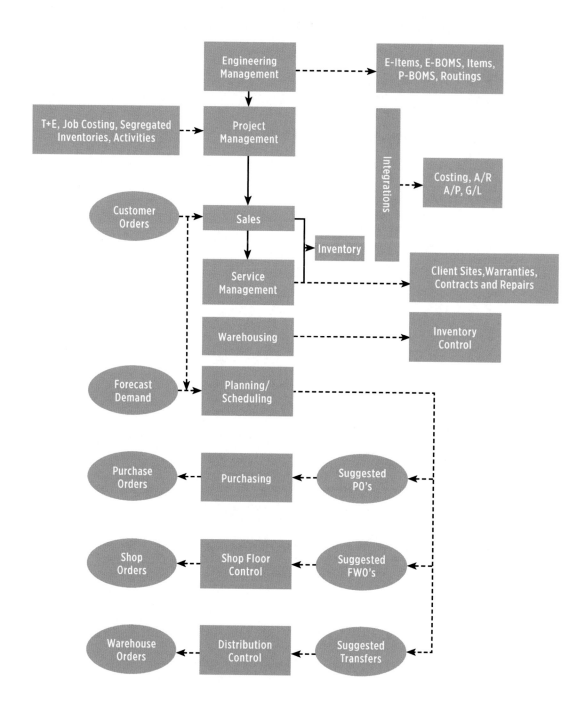

SERVICE MANAGEMENT

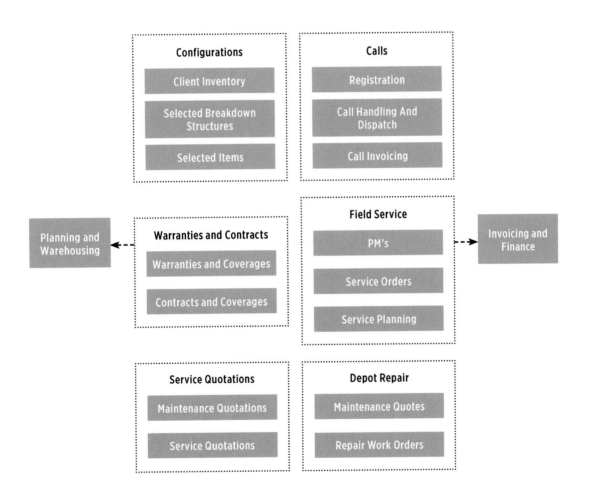

PROGRAM MANAGEMENT

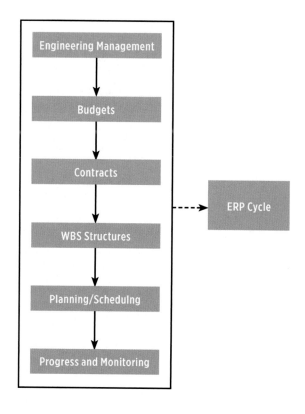

The Mapping Phase

Mapping refers to the documenting of the methods of conducting business in a company. These maps may take many shapes and forms but, ultimately, they are simply a collection of diagrams and flowcharts illustrating how to accomplish the day-to-day tasks of a department.

The Timeline

In the previous Core Team Training phase, the team members received sufficiently detailed functional training to enable them to adapt their current business process to the new software.

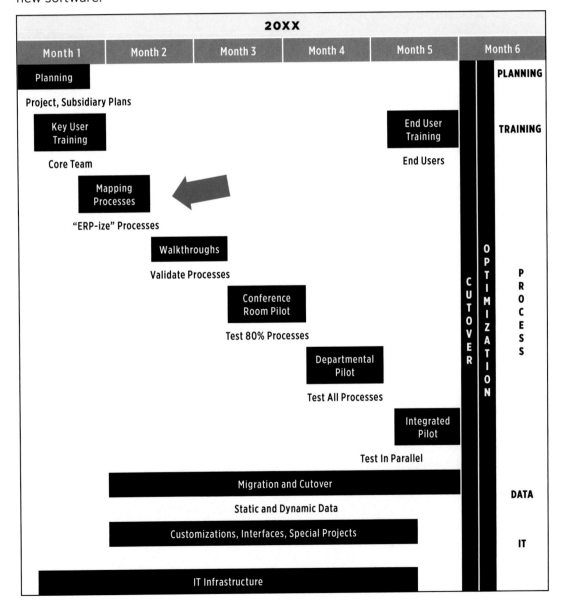

Following this mapping process, the theoretical approach will be presented to the rest of the core team during walkthroughs for validation and revision.

What this Phase Accomplishes

The Mapping Phase "ERP-izes" the existing business processes. This means that existing flowcharts, work instructions, and process documentation must be adapted to meet the demands of the new ERP application. Of course, if existing documentation is unavailable, it should first be created from scratch.

The blueprint white paper incorporates these adapted "to be" process mappings as the new departmental flows. This deliverable begins to take shape during this phase.

As the core team progresses through this new business process design, the many gaps and issues that surface are diligently logged. Also, end user skill levels are noted and corresponding change management training grids are assembled. In this manner, the end users' training requirements are developed and documented.

The Tasks

1) For each department, list the *existing* business scenarios

   For each identified scenario, identify whether it occurs frequently (and falls into the 80% category), or infrequently (and falls into the 20% category). Indicate whether the scenario is carried out manually or on the legacy system. Also indicate if it is to stay manual or be carried out on the new ERP system.

2) Review the existing business processes.

   If these "as is" flows are unavailable, they should be documented. This may require a department-by-department interview process. Not only will they be used by the consultants to learn the way business currently operates, but they will also be used as a benchmark against which the "to be" process flows will be compared.

3) Key users, with help from the outside consultants, must map the existing business processes onto the new software.

   These maps form the basis for the business model and flows published in the blueprint white papers. These are the "to be" flows and procedures.

4) As it becomes clear how the business will use the new software, employees' job functions may change. Refine the change management skills grid, training recommendations, resource, and cost estimates for all users whose jobs will change as a result of the ERP implementation.

5) Identify all gaps, issues, and problems, and assign them priorities; estimate the resolution effort; and assign a responsible key user. Review these with senior management.

6) Populate the ERP application setup and parameter files.

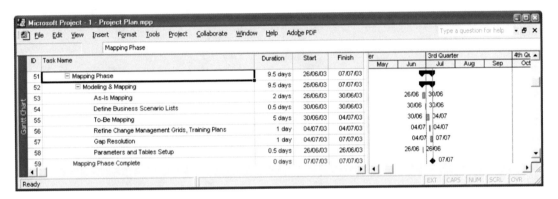

Figure 10. Mapping Tasks

The Deliverables

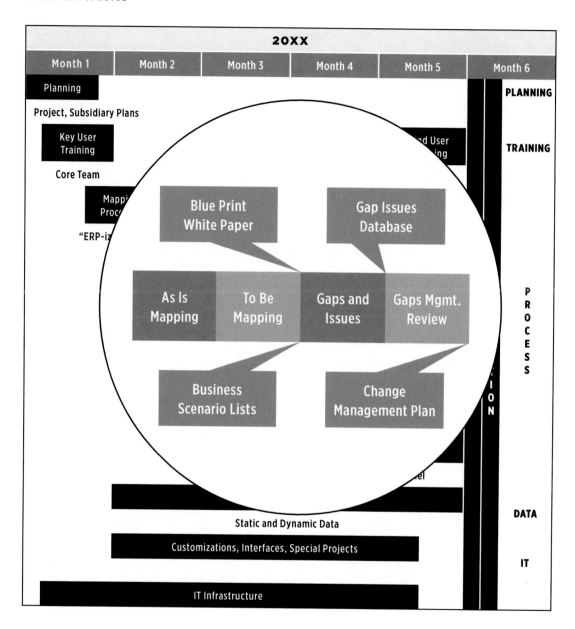

The Fourth Deliverable: Business Scenario Lists

A business scenario is a way of rigorously quantifying how a department conducts a specific business function. It is a label for a business process's unique series of detailed user instructions to get the work done. (The scenario script itself is discussed in the next session.)

For example, the entry of a cash-on-delivery (COD) sale involves user steps that differ significantly from those needed to enter a credit card sale. Each is a separate business scenario with its associated scenario script.

Avoid creating too many separate scenarios. A scenario's script may include decision points or branches. But, if a decision point leads into a highly divergent series of steps before it joins back up, it probably should be split into two separate business scenarios.

When all daily, weekly, monthly business processes are each covered by a scenario, the departmental scenario list is complete.

Next, identify how the scenario is currently handled. Determine whether the scenario will require the use of the new ERP application, be entirely manual, or require some other system or piece of software.

Then, identify each entry as an 80% or 20% scenario. 80% scenarios are those used frequently; they will be scripted and piloted during the coming Conference Room Pilot Phase. The 20% scenarios will be scripted and tested later during the Departmental Pilot Phase.

Although each company differs, the following are examples of this 80% - 20% division:

80% Scenario Examples

- A sale of standard product inventory
- The receipt of an inspectable inventory purchased item
- The receipt of a customer invoice payment

20% Scenario Examples

- A product return for warranty repair
- The purchase of an asset

Finally, indicate the status of the scenario's associated script. This helps in tracking how the core team is doing in completing its scripting mandate. Scripting statuses are:

1. Open: the scenario script has yet to be written
2. Written: the script has been written, but not tested nor reviewed
3. Reviewed: the script has been tested in the new ERP application
4. Complete: the review is done

Warehousing Business Scenario List						
	80% 20%	Work Instructions/ Script ID	White Paper	White Paper Completed	Script Completed	Reviewed & Verified
Warehousing Overview						
1 Overview	NA	NA	Warehousing Overview.doc	x	NA	NA
Outbound Process - 1						
1 Outbound Process - Sales	80	101	Warehousing Outbound Process.doc	x	x	
2 Outbound Process - Purchase Returns	20	102	Warehousing Outbound Process.doc			
3 Outbound Shortage Process - Sales	20	103	Warehousing Outbound Process.doc	x		
4 Outbound Process - Manufacturing	80	104	Warehousing Outbound Process.doc			
5 Outbound Shortage Process - Manufacturing	20	105	Warehousing Outbound Process.doc			
6 Outbound - Batch Process	20	106	Warehousing Outbound Process.doc	x	x	x
7 Outbound Exception Process	20	107	Warehousing Outbound Process.doc			
Inbound Process - 2						
1 Inbound Process - Purchasing	80	201	Warehousing Inbound Process.doc	x		
2 Inbound Process - Sales Returns	20	202	Warehousing Inbound Process.doc			
3 Indirect Receipts Process	80	203	Warehousing Inbound Process.doc	x	x	x
4 Inbound Process - Manufacturing	80	204	Warehousing Inbound Process.doc			
5 Direct Delivery PO Receipt Process	20	205	Warehousing Inbound Process.doc	x		
6 Inbound Exception Process	20	206	Warehousing Inbound Process.doc			
Inspection Process - 3						
1 Inspection Process - Purchase Receipts	80	301	Warehousing Purchase Inspection Process.doc			
2 In-process Inspection - Manufactruing	80	302	Warehousing In-Process Inspection.doc			
Assembly Process - 4						
1 Simple Assembly Process	80	401	Warehousing Assembly Process.doc	x		
Transfer Process - 5						
1 Item to Item Transfers	20	501	Warehousing Transfer Process.doc			
2 Manual Transfers	20	502	Warehousing Transfer Process.doc			
3 Manual Quick Transfers	20	503	Warehousing Transfer Process.doc	x		
4 In Transit Process	20	504	Warehousing Transfer Process.doc			
Exception/Correction Process - 6						
1 Order Correction Process	20	601	Warehousing Correction Process.doc			
Audit Process - 7						
1 Physical Inventory	20	701	Warehousing Adjustment Process.doc			
2 Consignment Audit	20	702	Warehousing Adjustment Process.doc			
3 Inventory Adjustment Process	80	703	Warehousing Adjustment Process.doc	x		
Repair and Exchange Process - 8						
1 RMA process	20	801	Warehousing RMA Process.doc			

Figure 11. Sample Business Scenario Matrix

When all daily, weekly, and monthly business processes are each covered by a scenario, the departmental scenario list is complete.

Figure 11 illustrates an Excel Business Scenario list for a sample Warehousing functional area.

The Fifth Deliverable: Blueprint White Papers

The blueprint white paper describes the business process of a specific functional area or department. It shows how this department fits in with the others, from a process and paper-flow perspective. Included are flow-charted process flows, selected screen snapshots, key-field values and descriptions.

The blueprint white paper evolves throughout the project cycle.

Figure 12 illustrates how this document improves as the core team's knowledge increases. This first draft of this document will be used in the walkthroughs presented in the next phase.

We cannot stress enough the importance of the blueprint white papers which are incorporated into the user documentation, long after cutover. Users and key-users alike will want to review how decisions were made to handle a particular business function.

A blueprint white paper sample table of contents follows:

1) Introduction

    Introduce the functional area and list any major deviation from existing departmental processes; highlight any item requiring particular emphasis.

2) General Overview

    Present the high-level view of the entire suite of the new ERP application modules; underline the relationships among the various components. This overview, repeated in every functional area's blueprint white paper, forms the basic overview in the end user documentation (Figure 13).

3) Functional Overview

    Highlight the ERP application modules of this particular functional area. Include their relationship to the other modules, and document the inputs to, and the outputs from these areas (Figure 14).

4) Functional Flows

Present the functional area's actual process maps (the "to be" flows). These reflect the day-to-day workflow showing the inputs to, and outputs from the processes, employee distribution lists, as well as other key data structures and decision points (Figure 15).

It is not the intent of this text to recommend a particular data flow diagram methodology.

5) Master Data, Session Screens and Key Fields

Define the important master data files and their contents. Provide screen snapshots, and highlight *only* key fields needing team consensus (Figure 16).

6) Reporting Requirements and Batch Processes

Define the reporting needs, reporting specifications, as well as batch processes that need scheduling.

7) Personalization and Customizations, Interfaces and Special Projects

Define the changes to the basic system as well as "bolt-on" interfaces that require programming.

8) Gaps and Issues

Record and prioritize gaps and issues in the document's final section. Alternatively, use a separate Gaps and Issues database.

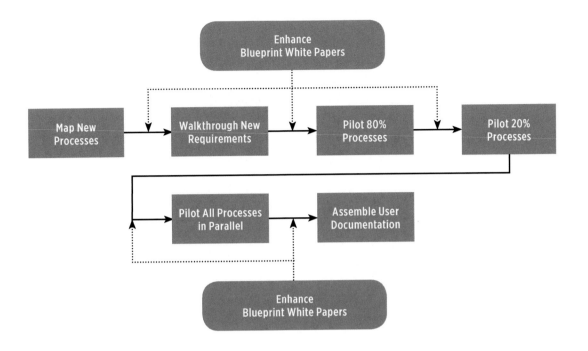

Figure 12. Blueprint White Papers Iteration Points
The blueprint white paper evolves throughout the project cycle as the core team's knowledge increases.

44

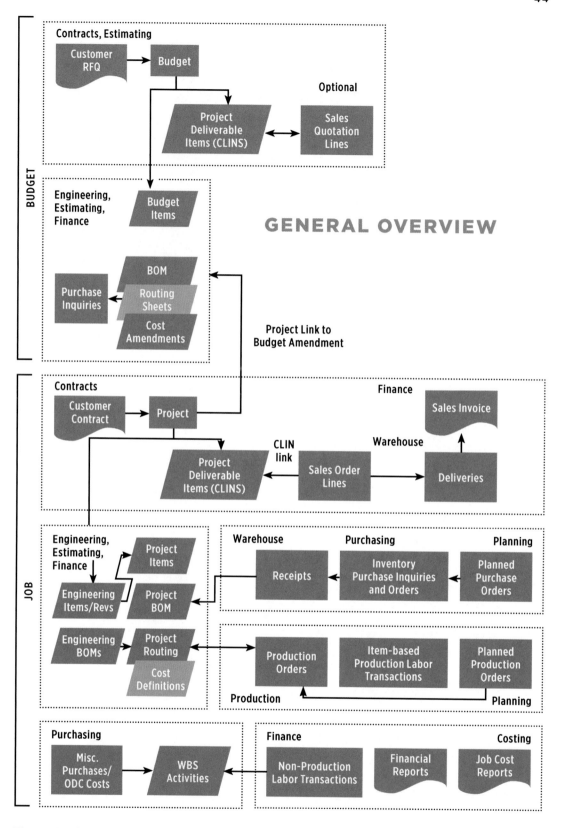

Figure 13. Blueprint White Paper General Overview

ENGINEERING OVERVIEW

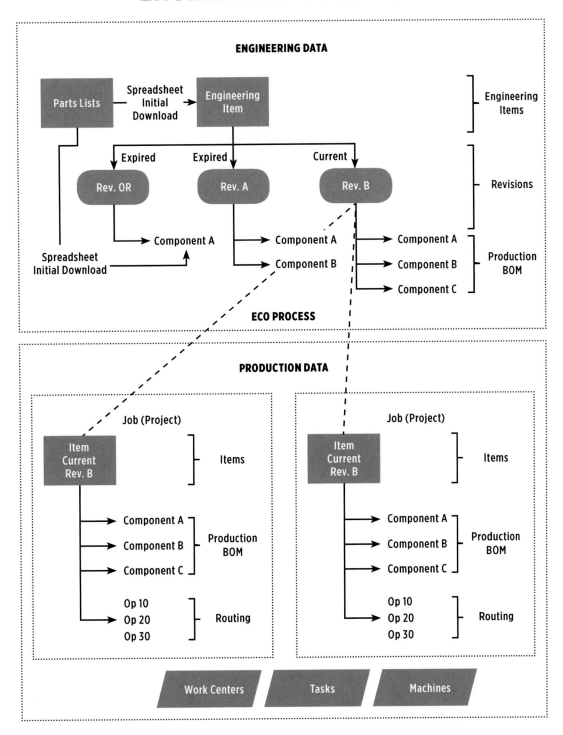

Figure 14. Blueprint White Paper General Overview

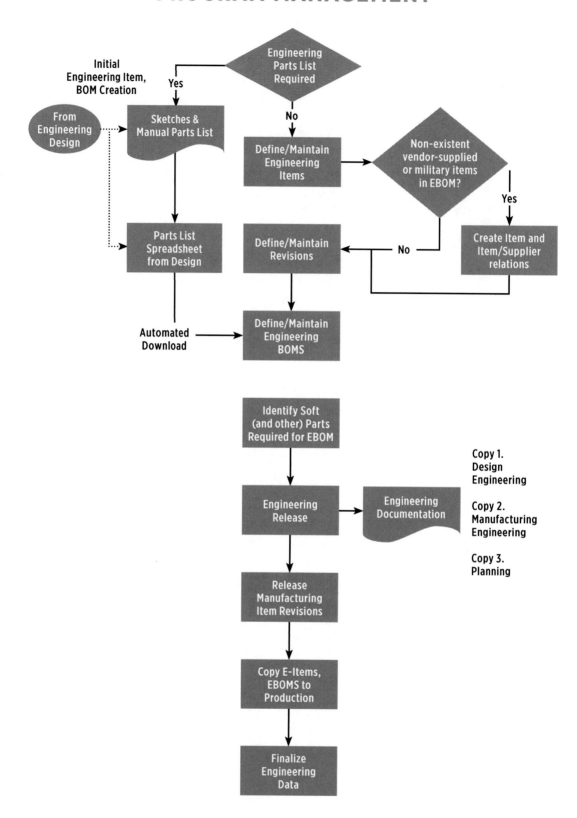

Figure 15. Blueprint White Paper Functional Flow

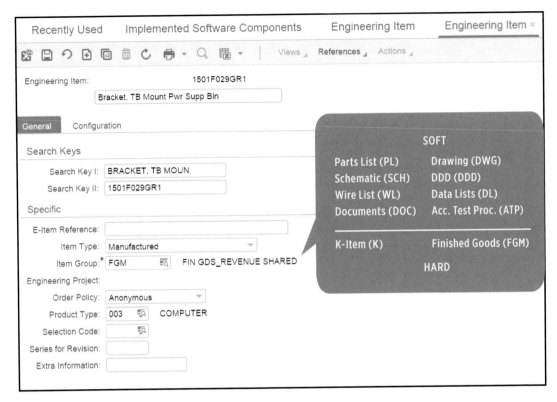

Figure 16. Blueprint White Paper Detailed Screen Snapshot

The Sixth Deliverable: The Gaps and Issues Database

From a "40,000-foot" view of the implementation strategy, the core team's primary objective, after mapping and testing the process flows, is to systematically close all gaps and issues.

However, as the project progresses from phase to phase, these gaps and issues accumulate with frightening speed. One way to manage the list effectively is to establish a scheme that prioritizes serious gaps and issues ahead of the less urgent ones.

Assign "high" priority to those that should be resolved before cutover. High priority gaps and issues lacking a closure plan will prevent us from cutting over to the new system; these are labeled as "show-stoppers."

Assign "medium" priority to those gaps and issues that require closing shortly after cutover.

Assign "low" priority to gaps and issues that are considered "nice to haves" but are not essential to the operation of the business.

It is evident that "high" priorities or "show-stopping" gaps and issues must be addressed first. No "show-stopper" should be on the list by the time the dynamic data entry begins during the Cutover Phase.

"Medium" priority gaps and issues usually relate to financial reports or processes which, by the nature of the implementation schedule, are not due until the first financial month-end after cutover.

For gaps and issues that requiring IT involvement, record the estimate to complete the development effort; assign a "sub-priority" from 1 to 10. This sub-priority indicates which IT tasks are more important than others.

As the project moves from the Departmental to the Integrated Piloting Phase, the team must diligently attack the gaps and issues daily. The weekly status meeting is a good place to ensure that teams are actively closing their "show stoppers" and "high" priority gaps and issues.

The risk management report must reflect all "show stopping" gaps and issues. It's the steering committee's responsibility to either help resolve these or decide whether the project should proceed if they persistently resist resolution.

A gap or issue is recorded as soon as it is brought up by a team member. If a work-around is identified or IT programming is required, enter a planned closed date. If the gap or issue is a "show-stopper", record it as such.

When a gap or issue is closed, do not delete it from the list, but record its closure and set its closed date.

Figure 17 and the reports that follow illustrate a Microsoft Excel format for tracking and reporting gaps and issues.

	**Pemeco Gaps and Issue Worksheet**

Report Date:
24-Jan-13

Department:
Gaps and Issues Consolidated Sheet

Key User:
Peter Gross

ID	Department	Description	Responsible Core Team	Responsible Consultant	Type of Issue	Issue Date	Due Date	Priority	Status	Notes
GA001	Engineering	Define ION workflows for New Items, Revised Items	Ben A, Steve P.		Process	January 7, 2013	January 31, 2013	1	Open	
GA002	Engineering	How to supply purchasing with manufactures pdf for new items	Ben A.		Process	January 7, 2013	January 31, 2013	1	Open	
GA003	Engineering	Do we need to make a new ECO database?	Ben A.	Felipe	Process	January 7, 2013	2/31/2013	2	Open	will be looked at after migration
GA004	Sales	Obtain the price books for all customers (migration)	Bonnie S.	Felipe/ Tony	Data	October 4, 2012	October 11, 2012	2	Open	
GA005	Sales	Determine the format of the sales acknowledgement	Bonnie S.	Nick Z.	Report	September 28, 2012	September 28, 2012	1	In Process	
GA006	Sales	Do away with XM,XP,XT, SPL (sales order as concession "SC"), W will be (Electronic)	Bonnie S.	Felipe / Tony	Custom	September 28, 2012	September 28, 2012	1	Open	Developing a new numbering system for all orders.
GA007	Sales	Automate the "Release to Invoicing" step for all order types.	Bonnie S.	Nick Z.	Process	October 4, 2012	October 4, 2012	1	Closed	
GA008	Sales	Implement cross reference table for legacy sales order number (migration)	Bonnie S.	Felipe / Tony	Custom	September 28, 2012	January 11, 2013	1	In Process	
GA009	Sales	Off Line Data bases (Service Information Sheet, Service Schedule, Field Service CSM)	Bonnie S.	Peter	Data	October 17, 2012	October 17, 2012	1	Open	
GA010	Sales	Reports /Forms used daily for Sales, Service, RMA process (Acknowledgement)	Bonnie S.	Nick Z.	Report	October 17, 2012	October 17, 2012	1	Open	
GA011	Sales	Tying serial number of finished goods to sales order (drives and starters)	Bonnie S.	Nick Z.	Data	October 17, 2012	October 17, 2012	1	Open	
GA012	Sales	Recording Commissions for Salesmen and Reps and Reporting	Bonnie S.	Felipe / Tony	Process	October 18, 2012	October 18, 2012	1	Open	
GA013	Sales	Credit Card Customer (enter as temporary cust. With ccd as payment type)	Bonnie S.	Celia K	Process	October 23, 2012	October 23, 2012	1	Open	
GA014	Sales	Mike's Rep / Sales request List (email)	Bonnie S./Steve P	Celia / Nick	Process	October 23, 2012	October 23, 2012	1	Open	
GA015	Sales	Shipping codes (UPSRED, UPSGRN, UPSRAM) Automated UPS System to integrate with LN	Bonnie, Troy, Tony	Felipe S.	Data	November 14, 2012	November 30, 2012	1	Open	
GA016	Sales	What is the process for adding or modifying addresses on the fly	Bonnie S./Steve P	Nick Z.	Process	November 29, 2012	November 30, 2012	1	Open	What is the process for adding or modifying addresses on the fly.
GA017	Sales	Need 3yr data for warranty info. & all VFD's	Bonnie S./Tony G.		Data	December 5, 2012	December 26, 2012	1	Open	Importing 3 yrs Data for Service Call Manager
GA018	Sales	Add a PO Field to the Call Screen for the tech to enter the PO if it is given to him. This needs to be a searchable field.	Bonnie S	Felipe / Tony	Data	December 13, 2012	January 15, 2013	1	Open	
GA019	Planning	No planning BOMs for MRP planning	Ben A	Peter	Process	November 26, 2012	December 21,2012	2	Open	need to speak to Ben A for BOM
GA020	Planning	Lack of forecast for SIOP	Sharilyn	Peter	Process	November 26, 2012	December 21,2012	2	Open	Need to speak to Mike for forecast.
GA021	Purchasing	Does BAAN completely Replace e-Procurement (a Benshaw Program?)	Larry T.	Nick	Process	December 5, 2012	December 26, 2012	1	Open	
GA022	Purchasing	How will PI Cycles be handled in BAAN?	Larry T.	Nick	Process	December 5, 2012	December 26, 2012	1	Open	
GA023	Purchasing	How is Drop Ship orders handled in BAAN between Sales & Purch	Larry T.	Nick	Process	December 5, 2012	December 26, 2012	2	Open	
GA024	Purchasing	The look & design of the Benshaw Purchase Order Form in BAAN	Larry T.	Nick	Custom	December 5, 2012	December 26, 2012	1	Open	
GA025	Purchasing	Item Part Numbers. Using current numbers or creating new?	Larry T.	Nick	Data	December 5, 2012	December 26, 2012	1	Closed	
GA026	Warehousing	External databases and integration	Rob V.	Nick	Custom	December 6, 2012	December 26, 2012	1	Open	
GA027	Warehousing	UPS world ship , FedEx software , LPS	Rob V.	Nick	Custom	December 6, 2012	December 26, 2012	1	Open	
GA028	Warehousing	Daily shipping report?	Rob V.	Nick	Reports	December 6, 2012	December 26, 2012	1	Open	
GA029	Warehousing	Bill of Lading , Packing slips, commercial invoices	Rob V.	Nick	Report	December 6, 2012	December 26, 2012	1	Open	
GA030	Warehousing	Barcoding systems / scanning guns	Rob V.	Nick	Process	December 6, 2012	December 26, 2012	2	Open	
GA031	Warehousing	A,B,C, Item definitions for cycle counts.	Rob V.	Nick	Data	December 6, 2012	December 26, 2012	2	Open	
GA032	Manufacturing	Additional documents printing - drawings, schematics, labels	MFG	Peter/Felipe	Process	December 18, 2012	January 18, 2013	2	Open	
GA033	Manufacturing	Planned production start date range filter capability	MFG	Peter/Felipe	Process	December 18, 2012	January 18, 2013	2	Open	
GA034	Manufacturing	Cycle times for kitting operations	MFG	Peter	Data	December 18, 2012	December 31, 2012	1	Open	
GA035	Manufacturing	List of ship to site Machine Shop and Wire Harness center	MFG	Peter	Report	December 18, 2012	December 31, 2012	2	Open	
GA036	Manufacturing	Owner post-initial cutover in May (Mfg Eng)	MFG	Peter	Policy	December 18, 2012	January 31, 2013	2	Open	
GA037	Quality	Document Attachments. We need to be able to attach Pictures, Office Documents, etc to everything.	Paul P.	Peter G.	Data	November 1, 2012	December 26, 2012	3	In Process	Peter was working with LN to get the feature working
GA038	Quality	Need Fields Added to the LN NCMR these fields need to be sortable and searchable- Rejecting Location (, Suppliers RMA Number, Disposition Comments	Paul P.	Nick Z.	Data	December 1, 2012	December 26, 2012	1	Closed	These fields are on our current NCM form and are used on each NCM
		Corrective Action Table in LN will not handle Benshaws requirements. Create a new									

Figure 17. Gap/Issue Entry Form

As the gaps and issues accumulate, manage the list effectively and prioritize serious gaps and issues ahead of the less urgent ones.

Open Gaps/Issues by Key User (Full List)

ID Resource	Title and Description	Gap Is	Entry Date	Plan Close

ARARAT T.

| 181 Francois P. | Re-visit Routing Revisions
Allan wants to keep track of routing
revision history, each with its set
of signatures. | New Session | 04/12/2002
Medium

Sub-Priority | 31/12/2002
Est: 1 day

2 |

ELAINE D.

| 182 Peter G. | Write on-hand quantity migration
import Migration script required. | Process | 04/12/2002
Hi
Sub-Priority | 31/12/2002
Est: 1 day
1 |

KELVIN K.

88	Sales Order Series Are sales order numbers to be unique by salesman?	Process	19/09/2002 Medium Sub-Priority	19/10/2002 Est: 0 days 0
94 Francois P.	MO Form Define and Print MO form as designed in access	Report	19/09/2002 Medium Sub-Priority	 Est: 2 days 2
96 Peter G.	Show PO Information on Quotes Show PO information on "Enter Quotation Results" report.	Report	19/09/2002 Low Sub-Priority	01/12/2002 Est: 5 days 3
191 Jack T.	Sales Acknowledgement form Adjust form printout according to specifications.	Customization	09/12/2002 Medium Sub-Priority	31/12/2002 Est: 1 days 1

ID Resource	Title and Description	Gap Is	Entry Date	Plan Close
223 Peter G.	Zoom to External Cross Reference Need zoom to External, not Internal, Cross Reference	Customization	20/12/2002 Medium Sub-Priority	31/12/2002 Est: 1 days 3
228 Peter G.	Drop ship PO's need manual linking Automated links need to be modified from time to time.	Customization	20/12/2002 Medium Sub-Priority	31/12/2002 Est: 1 days 3

Open Gaps/Issues by Key User

ID Resource	Title and Description	Gap Is	Entry Date	Plan Closed
KELVIN K.				
88	Sales Order Series Are sales order numbers to be unique by salesman?	Process	19/09/2002 Medium Sub-Priority	19/10/2002 Est: 0 days 0
94 Francois P.	MO Form Define and Print MO form as designed in access	Report	19/09/2002 Medium Sub-Priority	 Est: 2 days 2
96 Peter G.	Show PO Information on Quotes Show PO information on "Enter Quotation Results"	Report	19/09/2002 Low Sub-Priority	01/12/2002 Est: 5 days 3
191 Jack T.	Sales Acknowledgement form Adjust form printout according to spec.	Customization	09/12/2002 Medium Sub-Priority	31/12/2002 Est: 1 days 1
223 Peter G.	Zoom to External Cross Reference Need zoom to External, not Internal, Cross Reference Inquiry	Customization	20/12/2002 Medium Sub-Priority	31/12/2002 Est: 1 days 3
228 Peter G.	Drop ship PO's need manual linking Automated links need to be modified from time to time.	Customization	20/12/2002 Medium Sub-Priority	31/12/2002 Est: 1 3

Open Gaps/Issues by Functional Area

ID Resource	Title and Description	Gap Is	Entry Date	Plan Closed
KELVIN K. (Contract/Sales)				
88	Sales Order Series Are sales order numbers to be unique by salesman?	Process	19/09/2002 Medium Sub-Priority	19/10/2002 Est: 0 days 0
94 Francois P.	MO Form Define and Print MO form as designed in access	Report	19/09/2002 Medium Sub-Priority	Est: 2 days 2
96 Peter G.	Show PO Information on Quotes Show PO information on "Enter Quotation Results"	Report	19/09/2002 Low Sub-Priority	01/12/2002 Est: 5 days 3
191 Jack T.	Sales Acknowledgement form Adjust form printout according to specs.	Customization	09/12/2002 Medium Sub-Priority	31/12/2002 Est: 1 days 1
223 Peter G.	Zoom to External Cross Reference Need zoom to External, not Internal	Customization	20/12/2002 Medium Sub-Priority	31/12/2002 Est: 1 days 3
KELVIN K. (PURCHASING)				
228 Peter G.	Drop ship PO's need manual linking Automated links need to be modified	Customization	20/12/2002 Medium Sub-Priority	31/12/2002 Est: 1 days 3

The Seventh Deliverable: The Change Management Plan

1) Gap Analysis

    a) Task analysis worksheets
       i) One per department
    b) Analyze "to be" jobs and gaps from "as is"

2) Assign individuals to jobs

3) H-R Alignment

    a) Create new job descriptions
    b) Re-grade new jobs, if required
    c) Determine Transition Plan for displaced employees
    d) Begin recruiting process for new hires
    e) Meet with employees to be transitioned

4) Communication Plan

    a) Design
    b) Implement

5) Finalize Training Strategy

    a) In-house or outsourced
    b) Geographic constraints
    c) Physical locations and environments
    d) Develop training assessment

6) Training Budget

    a) CBT & purchased packages
    b) Travel
    c) Consulting time
    d) Materials & resources

7) Selected End User Training

    a) Identify participants
    b) Overview training program
        i) Training development
        ii) Training materials
        iii) Training logistics
        iv) Conduct training
            1) Introduction
            2) ERP Overview
            3) ERP Navigation
    c) Functional Training Program
        i) Training development
        ii) Finalize training for selected end users
        iii) Conduct training
    d) Conduct training effectiveness assessment
    e) Review selected end user training for effectiveness

8) General User Training
    a) Follow selected end user training path
    b) Conduct training effectiveness assessment

The Walkthrough Phase

Walkthroughs are formal presentations of the initial cut of the blueprint white paper's departmental overviews and flows.

The Timeline

In the previous Training and Mapping phases, the core team members received sufficiently detailed ERP application functional training to enable them to adapt their current business process to the new software. The Mapping Phase required them to develop the new business flows and procedures and prepare the blueprint white papers.

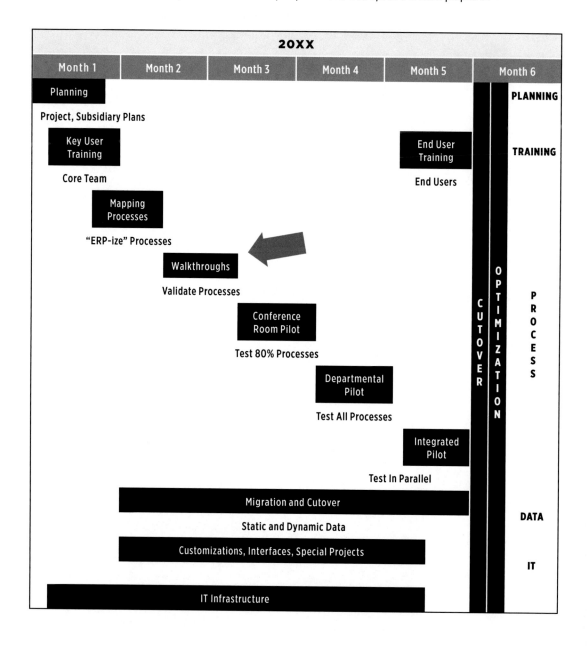

Following these presentations on how to do business with the new ERP software, three phases of intense pilot testing are conducted to prove out the concepts.

What this Phase Accomplishes

It's show and tell time for the core team. In the "to be" flows, the team designed how they are going to use the new software to do their work. Now, each functional area must demonstrate to the rest of the team how they've melded their department's business processes with the new ERP software.

This is done by walking the entire core team through the blueprint white paper's overviews and the "to be" flows using presentations of flow charts and application screen snap-shots. *You can avoid extending these presentations into marathons by presenting the ERP sessions' key fields and key field contents only!*

A second but equally important objective of the walkthrough presentations is to verify that the choices made by one functional area do not conflict with those made by the others. *These "integration points" are a fundamental reason the complete core team must be present throughout the entire series of walkthroughs.*

As presentations progress, gaps and issues are added to the ever-growing list. At the close of each presentation, the list of added gaps and issues is reviewed and priorities are applied to each. Each is assigned a responsible core team member and a closure date.

Microsoft PowerPoint is our tool of choice for most walkthroughs. However, others have used Microsoft Word, Excel, or other software they are comfortable with. Don't lose sight of the objective which is to show the rest of the team how you plan to use the new software to run your department, not how sexy the presentation can be made. Be sure that the time to prepare these walkthroughs does not become excessive.

A word of warning...

In decades of running ERP projects, we always remind the core team presenters that *walkthroughs are not functional module training sessions.* Do not waste precious time navigating from screen to screen and field to field, unless these are important processes and important key fields to highlight!

The Tasks

1) Prepare the departmental walkthroughs

    Using any appropriate software, prepare each department's proposed business flows under the new ERP software.

2) Present the departmental walkthroughs

    These processes must be approved by the entire core team. In this way, not only will each key user provide his or her input to the process being presented, but all integration points with other business processes will be identified and reviewed for correctness.

3) Refine the business model and flows in the blueprint white paper

    The white paper flow charts, screen snapshots, and instructions must be kept up-to-date reflecting the changes identified during the walkthrough process.

4) Update the open gaps and issues

    As new ones are identified, the gaps and issues data base must be updated.

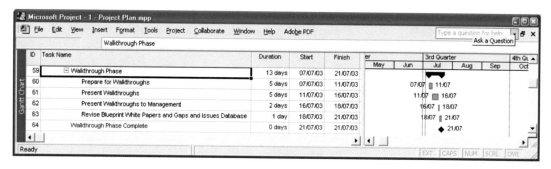

Figure 18. Walkthrough Tasks

The Deliverables

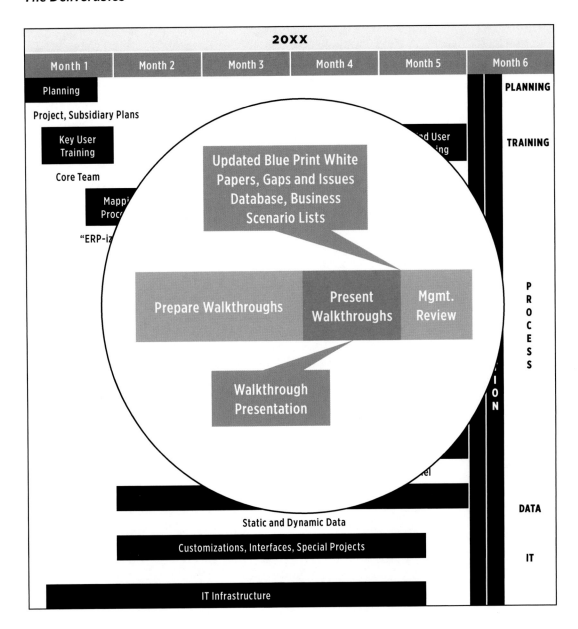

The Eighth Deliverable: The Walkthrough Presentation

Refer to Figure 14 and Figure 15 under the Blueprint White Paper section for samples of flow charts that are used in the walkthrough presentations.

Figure 19 below illustrates the use of a sample screen snapshot to highlight those key fields (and key fields only!) requiring discussion.

Sales Order Header

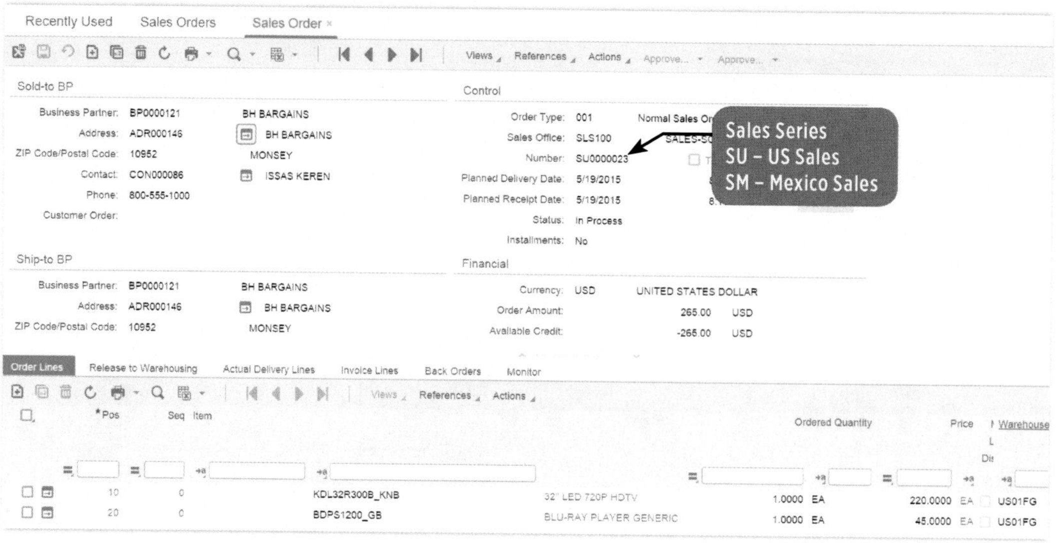

Figure 19. Screen Snapshot
Only key fields need highlighting!

Updated Blueprint White Papers, Business Scenario Lists, and Gaps and Issues Database

These documents are kept up-to-date with changes and additions identified during the walkthrough presentations.

The Conference Room Pilot Phase

The purpose of the Conference Room Pilot (CRP) Phase is to validate, in the new ERP application sessions, the theoretical business processes presented to the core team during the Walkthrough Phase.

The Timeline

The core team presented their conceptual view of the business processes during the prior Walkthrough Phase. The first series of actual tests on the software during this phase will be followed by more detailed testing in the coming Departmental Piloting Phase.

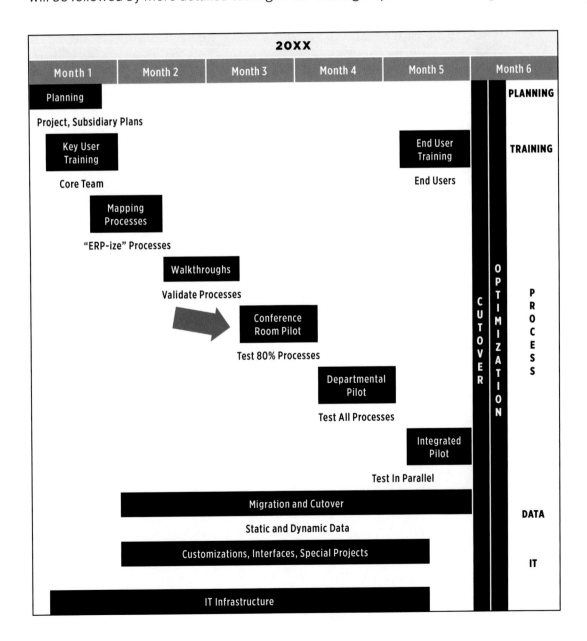

What this Phase Accomplishes

This phase tests the 80% most frequently used business scenarios, one functional area at a time. Here the entire core team participates directly, testing the ERP application software using a limited, manually created test-bed of data.

The Tasks

1) Review the system parameter settings.

    Results of the walkthroughs have most likely affected previous assumptions about how the system's parameters should be set. These must be reviewed and adjusted as required.

2) For each department, update the existing business scenario lists.

    Record any changes and omissions highlighted by the walkthroughs in the business scenarios.

3) For each department's 80% scenario, prepare a script detailing the steps needed to accomplish the scenario in the new ERP environment.

4) Validate the 80% business scenarios and scripts for correctness according to the ERP system, using staged test data in front of all key users.

5) Refine the business model and flows in the blueprint white paper.

6) As major adjustments to the flows occur, a second run-through of the entire process may be required. Once complete, an end-to-end pilot which tracks single transactions through all departments (e.g., an order-to-cash process) should be conducted to test all department-to-department hand-off's or integration points.

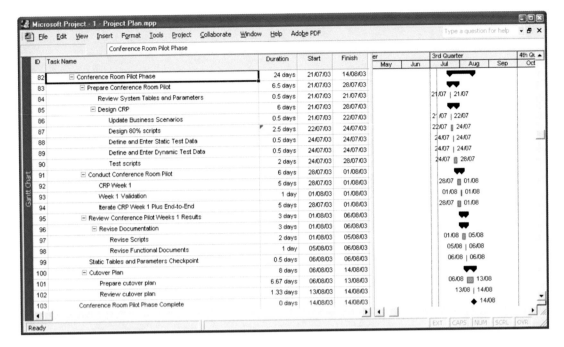

Figure 20. Conference Room Pilot Tasks

The Deliverables

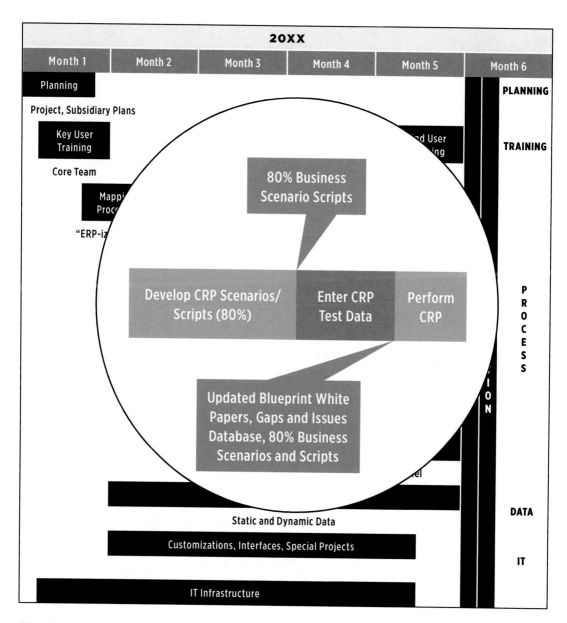

The Ninth Deliverable: The 80% Scenario Scripts

The business scenario lists developed earlier should cover all the business processes of each department. If we can define the recipe of user steps for every scenario, the collection of these scenarios and scripts will represent a significant portion of the users' cookbook of instructions. This will go a long way in documenting each department's day-to-day functions.

A script is nothing more than a set of user instructions detailed to the field level of the ERP application sessions.

Note that these scripts differ from a standard, generic ERP user instruction manual since they are specific to the company's way of using the software to conduct its business.

The following pages illustrate a sample script for the entry of a standard commercial sales order in a discrete manufacturing company.

Updated Blueprint White Papers, Gaps and Issues Database, and the 80% Business Scenarios and Scripts
These documents are kept up-to-date with changes and additions identified during the piloting.

SCRIPT SAL-23

Standard Commercial Sales Order

Sold To:	10-33 Fire Equipment Company BP C00000016
Invoice To:	10-33 Fire Equipment Company BP C00000016
Pay By:	10-33 Fire Equipment Company BP C00000016
Ship To:	XYZ Fire Department BP C00000017

Purchase Order number: 01-1001
Delivery: 1/25/01 UPS Ground
Terms: FOB Pine Brook, Prepaid

Item	Part Number	Description	Qty.	Price
1	KIT1	P-16 Kitty Hawk	1	$8,650.00

SCRIPT SAL-023
SALES ORDER ENTRY PROCESS

Step Taken (What I am going to do)	Step	Data Input Requirements	Session (Where I am going to do it)	Notes
Receive and Review Purchase Order		None required – review PO according to IAW departmental guidelines	NA	
Navigate to Sales Order entry screen		From Sales dropdown, select Sales Orders, Sales Orders, Sales Orders	Menu browser	

Master Data ⌄   CRM ⌄   **Sales ▾**   Project ⌄   Planning ⌄   Manufacturing ⌄   Procurement ⌄   Warehousing ⌄

Recently Used

Customer 360		
Sales Master Data ▸		
Sales Quotations ▸		
Sales Orders ▸	Sales Orders ▸	Sales Orders
Sales Contracts ▸	Sales Order Tracking ▸	Sales Order
Sales Schedules ▸	Back Orders ▸	Sales Delivery Workbench
Margin Control ▸	Installments ▸	Sales Order Line Delivery Overview
Consumption Handling ▸	Sales Order Blocking ▸	Sales Order History View
Retrobilling ▸	Sales Order Priorities ▸	Release Manual Activities Sales Order Lines
Commissions and Rebates ▸	Copy Sales Orders ▸	
Statistics ▸		
Sales Parameters ▸		

SALES ORDER HEADER ENTRY

Step Taken (What I am going to do)	Step	Data Input Requirements	Session (Where I am going to do it)	Notes
Create New Sales Order		Select [Insert] on screen to create new Sales Order Header	Sales Orders (tdsls4100m000)	

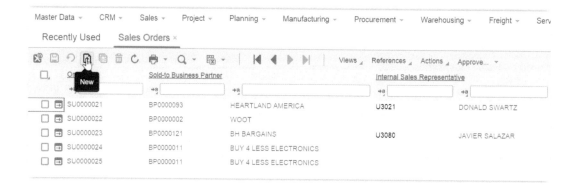

SALES ORDER HEADER ENTRY (CONT.)

Step Taken (What I am going to do)	Step	Data Input Requirements	Session (Where I am going to do it)	Notes
GENERAL TAB Enter Sales Order Header Data	1	SOLD-TO BP (Drill down to Sold-To Business Partners)	Sales Order Header	10-33 Fire Equip. C00000016
	2	SHIP-TO BP (Drill down to obtain correct address within client family)		10-33 Fire Equip. C00000017
[Sales Order Block]	3	SALES ORDER TYPE		012–Commercial Production
	4	SALES ORDER NUMBER Enter SU for US Sales Orders SM for Mexico Sales Orders		SU (Last 7 digits will be next available number)
	5	PLANNED DELIVERY DATE Enter planned "off-dock" date		01/25/15
	6	PLANNED RECEIPT DATE Enter customer's contract delivery date		02/02/15
➤ Advance to Sales Order Lines	7			

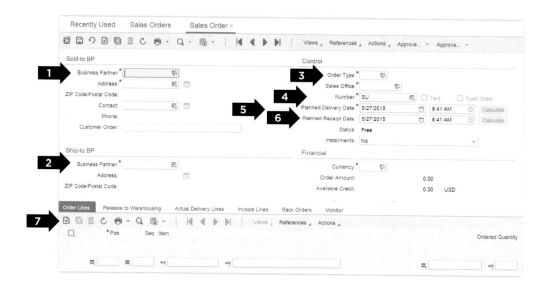

SALES ORDER LINE ENTRY

Step Taken (What I am going to do)	Step	Data Input Requirements	Session (Where I am going to do it)	Notes
Enter Sales Order Line Data		Click [Insert] Box to Open Sales Line Entry	Sales Order Lines	

Recently Used   Sales Orders   Sales Order ×   Sales Order Lines

Views ⌐ References ⌐ Actions ⌐ Approve... ▼ Approve... ▼

Sold-to BP

Business Partner:	BP0000011	BUY 4 LESS ELECTRONICS
Address:*	ADR000007	HISENSE
ZIP Code/Postal Code:	30024	SUWANEE
Contact:		
Phone:		
Customer Order:		

Control

Order Type:	001	Normal Sales Order - No Ack
Sales Office:	SLS100	SALES-SOHNEN ENTERPRISES, INC.
Number:	SU0000025	☐ Text ☐ Rush Order
Planned Delivery Date:*	5/25/2015	1:00 PM Calculate
Planned Receipt Date:	5/25/2015	1:00 PM Calculate
Status:	Modified	
Installments:	No	

Ship-to BP

Business Partner:*	BP0000011	BUY 4 LESS ELECTRONICS
Address:	ADR000013	BUY 4 LESS ELECTRONICS
ZIP Code/Postal Code:	80202	DENVER

Financial

Currency:	USD	UNITED STATES DOLLAR
Order Amount:		0.00 USD
Available Credit:		-4320.00 USD

Order Lines   Release to Warehousing   Actual Delivery Lines   Invoice Lines   Back Orders   Monitor

Views ⌐ References ⌐ Actions ⌐

New Order Line

		*Pos	Seq	Item		Ordered Quantity
☐		10	0	GM_0109	GM PALLET BLACK AND DE(	1.0000 EA
☑		20	0	1000231124	REMOTE(RM-YD092)BLK.SO	1.0000 PCS

SALES ORDER LINE ENTRY (CONT.)

Step Taken (What I am going to do)	Step	Data Input Requirements	Session (Where I am going to do it)	Notes
	1	ITEM		KIT1
	2	ORDERED QUANTITY		1
	3	PRICE ENTER UNIT PRICE (DEFAULTS)		$8650
	4	STANDARD DESCRIPTION Check box		✓
	5	SAVE AND EXIT		[All Other Tabs use defaults]

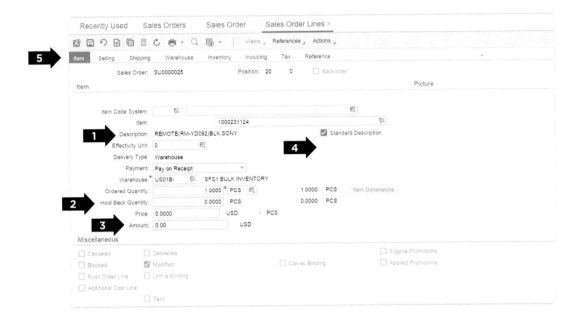

Updated Blueprint White Papers, Business Scenario Lists, 80% Business Scenario Scripts, Gaps and Issues Database

These documents are kept up-to-date with changes and additions identified during the walkthrough presentations.

The Departmental Pilot Phase

The Departmental Pilot (DP) introduces a new team of users to the testing process. They will re-validate the existing 80% business scenario scripts; they will also write and test the 20% scripts.

The Timetable

Once this phase is completed, the fully tested departmental business processes will be validated in an integrated fashion during the Integrated Piloting Phase.

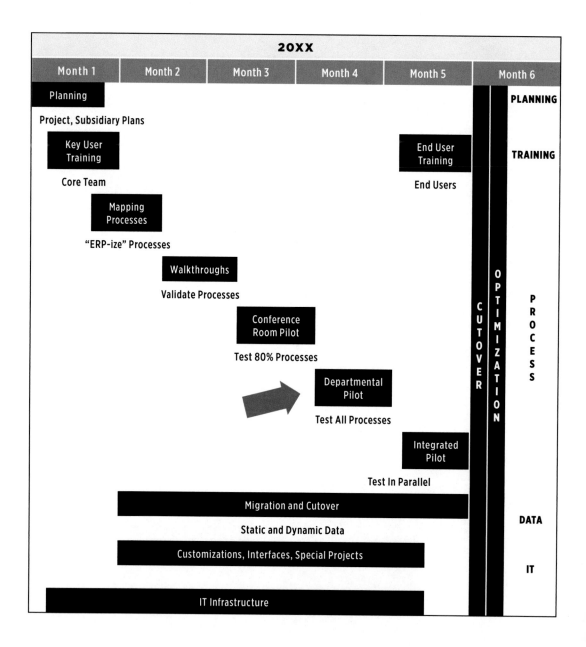

What this Phase Accomplishes

Since the key users' view of the day-to-day running of their departments is usually at a macro level, a group of selected end users with an intimate knowledge of daily processes is brought in to the departmental piloting cycle. Each selected end user must commit a minimum of 25% of their time — ideally a fixed number of hours per day (e.g., 2 p.m. – 4 p.m.).

The DP, the longest of the piloting phases, is conducted in each department in a decentralized fashion.

The first task is for the key users to train these users using courses similar to those presented to the core team. Moreover, the most recent flows in the evolving blueprint white papers now need to be integrated into the detailed functional training sessions.

DP must emphasize the "day-in-the-life" business processes. Up to this point in the piloting, much testing may have been done without regard for the daily (weekly, etc.) workflow of the end user. The DP teams must now focus on their daily schedules, to an hour-by-hour level of detail.

The final DP step is to test the scripts with a fully loaded, migrated database. This will validate the size and scalability of the hardware and system software. In addition, the customizations, interfaces, special projects and forms should now be available for preliminary testing.

The test data for DP resides in the Sandbox Company, previously used for the CRP. This data should be refreshed at the start of the DP.

Using the planning department as an example, the DP should provide answers to the following typical questions:

- Before I arrive in the morning, what is run at night to generate my MRP plan?
- At 8:00 a.m., what planning report do I need on my desk?
- At 9:00 a.m., what screens do I access to expedite and schedule my shop orders?
- What do I hand off to purchasing to execute my purchase plan?

At the conclusion of the DP, the numbers of gaps and issues must be decreasing. There should be no "show-stopping" issues without a pre-cutover closure plan.

System parameters and settings must be updated to reflect changes identified during this piloting process.

The Tasks

1) Train selected end users in a fashion similar to the core team training

   a) ERP overview training (all)
   b) ERP navigation training (all)
   c) ERP functional module training
   d) Business model (flows) functional training

2) Train the selected end users on the department's 80% scenarios and scripts

3) Key end-users and selected end-users prepare the remaining departmental 20% scenarios

4) Validate 80% and 20% business scenarios and scripts for correctness and volume constraints on the ERP system using static migrated data

5) Refine business model and blueprint white papers

6) Resolve all show-stopping gaps

7) Checkpoint ERP tables and parameters

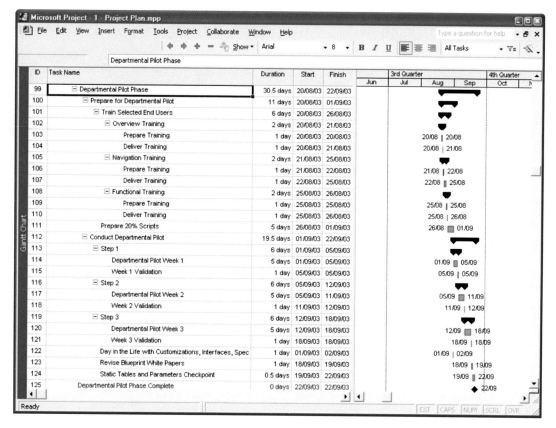

Figure 21. Department Pilot Tasks

The Deliverables

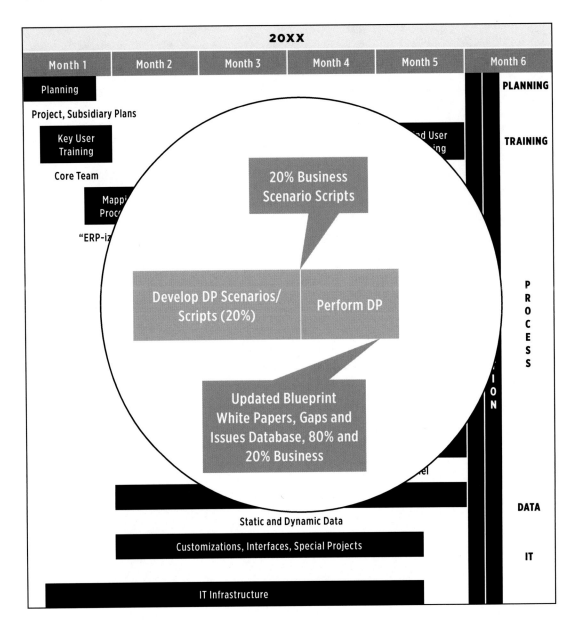

The Tenth Deliverable: The 20% Scenario Scripts
Please refer to the previous section for details and samples of the scenario scripts.

Updated Blueprint White Papers, Business Scenario Lists, 80% and 20% Business
Scenario Scripts, Gaps and Issues Database
These documents are kept up-to-date with changes and additions identified during the
walkthrough presentations.

The Integrated Pilot Phase

The Integrated Piloting (IP) is the final test phase before cutting over to the production system. All business processes are tested for accuracy across departmental lines.

The Timetable

The business processes were fully tested in a departmentalized fashion in the previous phase. Before the cutover to production, this final series of tests validates the full, integrated suite of procedures across the enterprise.

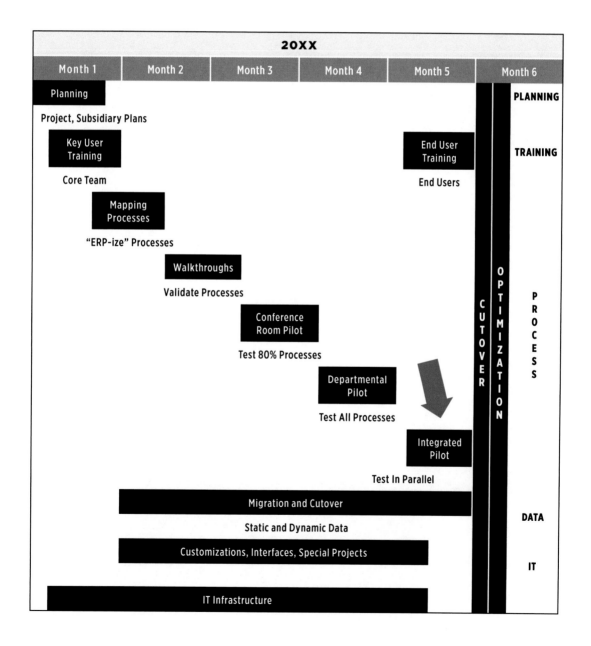

What this Phase Accomplishes

Selected legacy transactions are chosen to be input into the new ERP application. Outputs are compared for correctness and consistency. The IP brings together the new processes, the IT customizations, the interfaces and special projects in an integrated environment for the first time. The users' desktops, system security parameters, special printers, data collection systems, etc., must also be ready for testing during the IP Phase.

The IP requires significant preparation and coordination. Data must be migrated into a test bed and a representative subset of departmental legacy transactions chosen. Blocks of selected end-user time must be set aside for this pilot. Often, it is conducted on the user's screens and the system printers themselves. If this is not feasible, it can be conducted in a conference room environment. *The integrated piloting concept must use a "day-in-the-life" (not transactional) approach where the "hand-offs" between the many departments are carried out.*

Although not reflected in the project schedule, the volume of legacy data to be converted may cause an overlap of the IP with the long-term dynamic data entry tasks of the Cutover Phase.

The user documentation is updated to reflect any last-minute changes.

The Tasks

1) Finalize and generate user roles, desktops, and security for all functional areas.

2) (Optional) Re-migrate the static data files into the test environment.

3) Simulate live processing at user stations in the test system using actual legacy transaction inputs and "day-in-the-life" processes. To make the tests conclusive, choose a sufficient number of transactions and make them representative of a wide range of legacy functions.

Refine the business model and flows in the blueprint white paper.

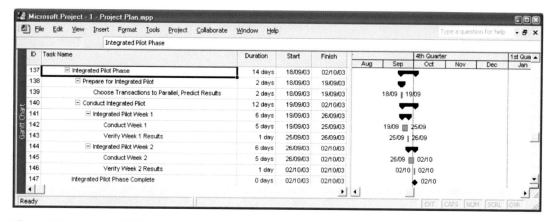

Figure 22. Integrated Pilot Tasks

The Deliverables

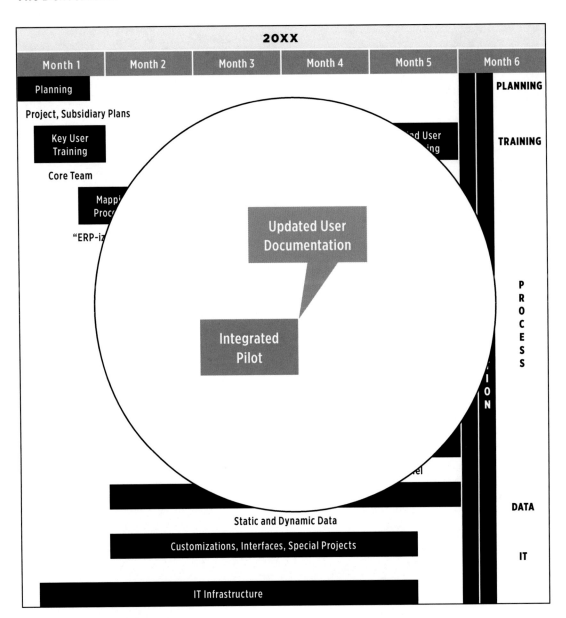

Updated User Documentation

These documents are kept up-to-date with changes and additions identified during the piloting.

The End User Training Phase

End users are any other users that require training on the system. These employees must attend the same battery of training sessions as the other team members.

The Timetable

The Change Management Skills Grid and End User Training Plan were developed during the Mapping Phase and, through the subsequent phases, kept up-to-date. This phase must ensure the end users are sufficiently trained prior to the final cutover weekend.

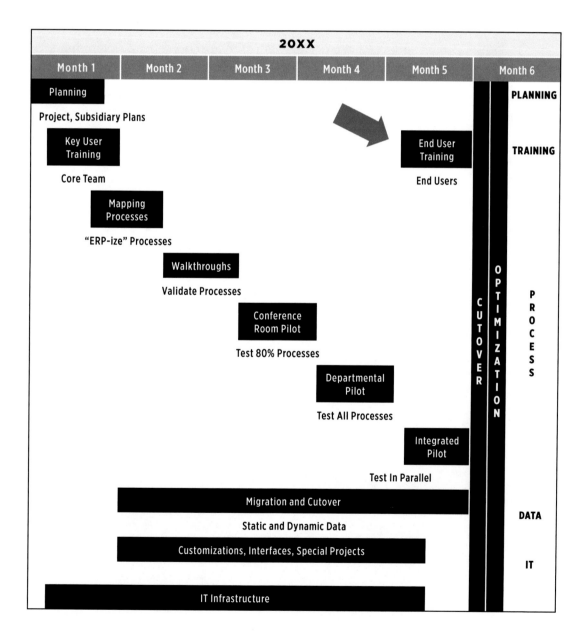

What This Phase Accomplishes

The training plan has detailed the course content, the delivery timetable, the trainers, and the geographic and logistic course-delivery constraints.

It is the core team members themselves who deliver the training to the end users. This training closely resembles that which they themselves received at the project's outset. However, the application-specific overview and the detailed flow training are tailored to the processes contained in the user documentation.

The Tasks

1) Train selected end users in a fashion similar to the core team training

    a) ERP overview training (all)
    b) ERP navigation training (all)
    c) ERP functional module training
    d) Business model (flows) functional training

2) Evaluate the training results

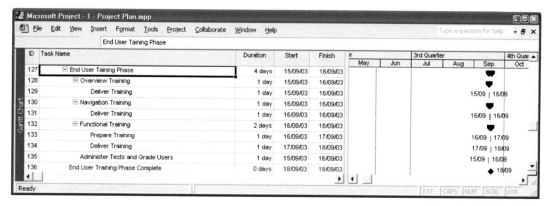

Figure 23. End User Training Tasks

The Deliverables

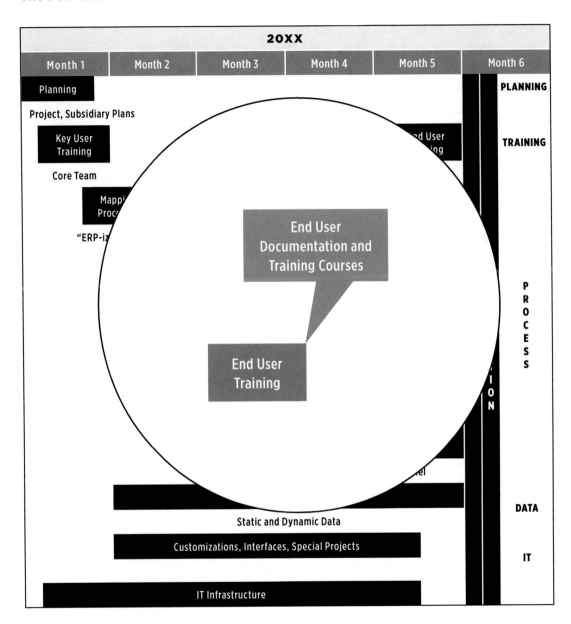

The Eleventh Deliverable: User Documentation

The User Documentation should be viewed as the business process bible for each department. A new hire should be able to pick it up and learn all about the department's operations. And, of course, the user documentation also serves as a reference guide to anyone needing to refresh their memory on how to do his or her day-to-day job.

The deliverables were developed in a specific sequence to enable simple re-assembly into a binder of documents that were prepared during the implementation cycle.

For each functional area, the binder contains the following:

1. The Blueprint White Paper
2. The Business Scenario List
3. The Scenario Scripts
4. Annex 1 – The Overview Training Course
5. Annex 2 – The Navigation Training Course
6. Annex 3 – The Functional Training Course

The Twelfth Deliverable: User Training Courses

1) Overview Training

   The Fundamentals training course is adapted to the end user's knowledge level and covers general ERP theory. This is followed by application-specific training which covers the ERP application software modules and their inter-dependencies.

1) Navigation Training

   Navigation training courses provide application-specific instruction on the manipulation of fields and the navigation of the screens, forms, and sessions.

2) Functional Training

   Functional training courses teach the functional details of the application modules.

3) Business Scenario and Script training

   Each business scenario and related script is covered in detail.

4) The overview and navigation courses followed by the core team and selected end users should suffice for the end users. The functional courses, once again, will require additional modification to reflect the ongoing workflow changes identified during piloting.

The Migration Phase

Bad data is one of the three primary causes of ERP project failures. *Bad data cannot be tolerated.* In other words, "Garbage in, garbage out."

Migration refers to the automatic or manual transfer and cleansing of any existing data into the files needed by the new ERP application to function. Sources for this data include the legacy system files, manual documents, spreadsheets, and so on.

The Timetable

The Migration Phase starts immediately after the core team is trained on the new ERP

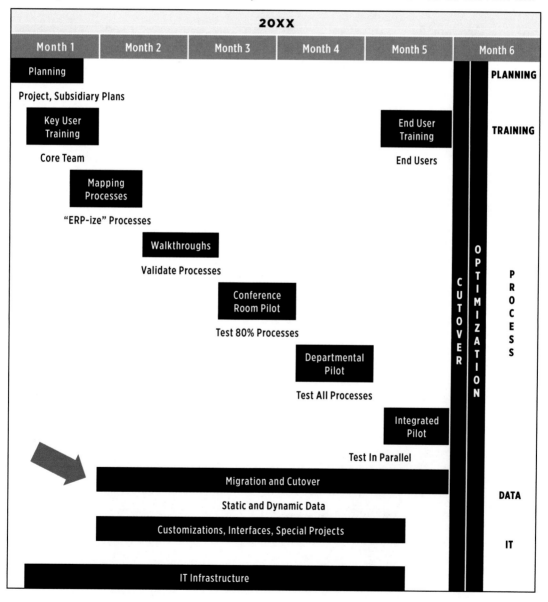

application and continues right through the final cutover weekend. It functions in parallel with the process phases of the implementation.

What This Phase Accomplishes

At this point, two definitions are required:

a) **Static data** is data that has a tendency to change infrequently. It can be migrated early in the implementation cycle to ease any last-minute rush, and to provide test data for piloting.

b) **Dynamic data** is data that is volatile. It should be migrated at the last possible moment to reflect its most up-to-date state.

Some examples of static data include:

Tables, Parameters, and Master Data

- Logistic, finance, and tax tables
- Application module parameters
- Non-volatile master data (order types, numbering series, user profiles, etc.)

Operating Data

- Business partners (customers, suppliers, etc.)
- Engineering master items, coding systems
- Engineering bills of materials, routings, etc.
- Employees
- Price books

Financial Data

- Charts of account
- Inter-module integrations (mappings)
- Financial budgets

Some examples of dynamic data include:

Operating Data

- Sales orders
- Purchase orders
- Shop work orders
- Inventory balances

Financial Data

- A/R open items
- A/P open items
- G/L balances
- Cost data

Dynamic data can be further sub-divided into:

- **Long-term** dynamic data
-
- **Short-term** dynamic data

Long-term dynamic data are those elements that are not expected to change between the time they are entered and the date the new system goes live. Examples are: sales and purchase order lines not scheduled for delivery or receipt until after cutover; shop work orders not scheduled to start before cutover.

Short-term dynamic data are those elements likely to change up to the final moments before final cutover. Examples include: sales order lines scheduled to ship before cutover; invoices with pending payments; costing files; and inventory balances.

Data cleansing strategies

The new system should start with a clean bill of health for all data. The data to be transferred to the new ERP application must be reviewed carefully to correct errors and redundancies. Any unnecessary or duplicate information must be removed.

There are at least three opportunities to cleanse data:

1. It can be cleansed prior to migration, using existing sessions on the legacy system itself.

2. It can be downloaded into an intermediate format (such as an Excel spreadsheet or Access database) and cleansed there before being transferred into the new ERP application files.

3. It can be cleansed after migration using the new ERP application sessions themselves.

Automated and manual conversion

Programs can be written to migrate data with little or no user intervention. The volume of data to be transferred and the compatibility of the legacy and the new ERP application record elements often determine whether fully automated routines can be used.

When coding automated migration routines, programmers must use rules to create structurally correct ERP application records that won't jeopardize the integrity of the file system.

A full re-migration of the static data can occur as often as necessary as long as no related operating data has been entered in the interim.

Maintaining data correctness

By definition, static data is cleansed and migrated to the ERP application files early in the final phases of the implementation cycle. Dynamic data is migrated as late as possible. However, the legacy system will still be operating, and some synchronization between the data in legacy and the newly migrated data must occur from that point forward. Any changes, additions, or deletions of records must be entered in duplicate to reflect the data correctly in both versions.

Identifying long and short-term dynamic data

Clearly, it will be necessary to get a head start on entering dynamic data since the last days before cutover will be exceedingly busy. This dynamic data must be prioritized to indicate which elements are long-term and which are short-term. The long-term dynamic data, by definition, should be entered before the short-term data since its contents are not as susceptible to change.

Print reports from the legacy system to list the data elements that should not change between their entry into the new ERP system and the cutover date; these should be ordered from most- to least-current date sequence. If these are available from the legacy system, include sufficient details to enable complete data entry from this report. If unavailable, print the location of the source documents.

Automated data migrations of both long- and short-term dynamic data need not be run as early in the cycle. An exception to this rule is for high-volume data that may force the migration routines to run over several days or weeks.

As the final cutover approaches, ship, receive, and close as many transactions as possible in the legacy systems to minimize the last-minute, short-term data entry. Once the legacy system is closed, enter the final open legacy data elements into the new ERP application. Print the legacy and new ERP application backlog reports to crosscheck and balance the two systems.

Attrition and converted data sets

It may be beneficial to leave some data on the legacy system to disappear by attrition depending on the volume of data to convert, its longevity, the compatibility of the legacy and the new ERP application file definitions and the ease of maintaining two systems and two sets of business procedures.

The Tasks

1) Prepare the Migration Plan

2) Execute the Migration Plan

Populate the static data files in the test environment for the departmental and integrated pilots and in the production files for final cutover
Populate any long- and short-term dynamic data files prior to cutover.

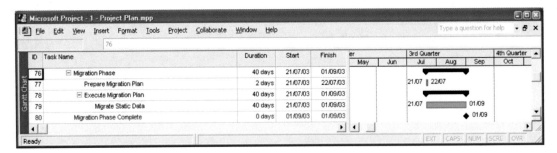

Figure 24. Migration Tasks

The Deliverables

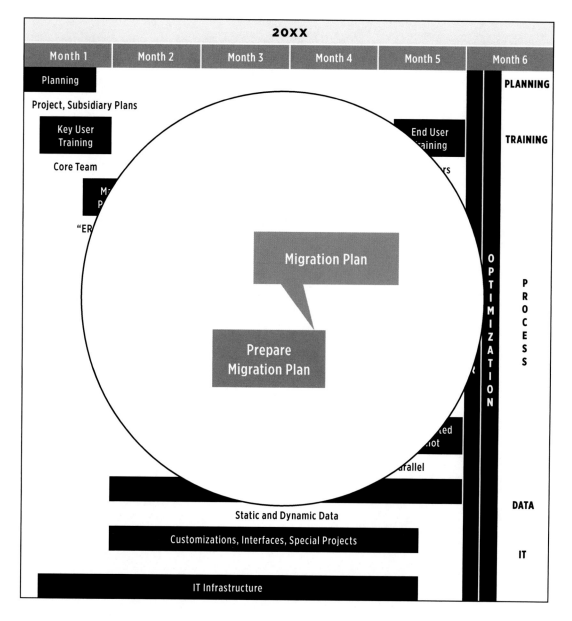

The Thirteenth Deliverable: The Migration Plan

A sample Migration Plan Table of Contents includes:

a) Introduction
b) Sandbox, QA, and Production Companies

c) Migration Strategies

(1) General approach

1. Locating data
2. Data cleansing
3. Maintaining correctness
4. Automatic versus manual migration

(2) Static data

1) Parameters, master data
2) Business partners
3) Item master records
4) Employees
5) Bills of materials
6) Routings

(1) Dynamic data

1) Open sales order lines
2) Open purchase order lines
3) Open work order lines
4) Open service order lines
5) Open warehouse order lines
6) Inventory details
7) Finance open items
8) Costing elements
9) Trial balance opening entries

d) Timetables

(1) Static data
(2) Long-term dynamic data
(3) Short-term dynamic data

The migration plan must address the methods used to transfer data into the new ERP application files.

Migration Strategies

a) Identify each ERP Application file to be populated, its source, and specify when it is to be done: early for static data; later for long-term dynamic data; or at the last minute for short-term dynamic data.

b) Determine the method of cleansing the data
   (1) In the legacy system
   (2) In an intermediate file using a secondary process such as Excel, Access, etc.
   (3) In the new ERP Application sessions, once transferred

c) Choose the migration method: automatic or manual entry

d) Determine the method to keep the data current in the Production Company
   (1) Re-migration
   (2) Dual legacy and ERP application manual update

Track the migration progress in a tabular matrix such as the one presented below.

Table	Original Data Source	Type	Cleanse Method	Migration Method	Production Update Method
Manufacturing Parameters	MAN	Static	MAN	MAN	Manual
Work Centers	LEG	Static	LEG	AUTO	Re-migrate
Machines	LEG	Static	ERP	MAN	Dual Manual
Bills of Material	LEG	Static	LEG	AUTO	Re-migrate
Routings	LEG	Static	LEG	AUTO	Re-migrate
Orders	LEG	Dynamic	LEG	MAN	Dual Manual

Sandbox, QA, and Production Companies

The IT environment for the ERP implementation must adhere to a set of quality control standards established at the project outset. With this methodology, we do not intend to suggest a single approach. Based on what has worked well in the past, we propose working within a simple framework of one or more instances or companies: Virgin, Sandbox, Development, Quality Assurance (QA), Migration, and Production.

The Virgin Companies preserve the parameters and master data settings and are maintained and kept up-to-date as any changes to these files occur. All the pilot testing occurs in the Sandbox Companies, and the quantity of these Sandboxes varies according to the piloting needs. Sandbox Companies are refreshed (as required from the Virgin Companies) via data entry or using automated migration routines.

ERP application version patches and IT-developed customizations are tested in Development Companies and selectively moved to QA Companies for pre-production testing and quality assurance. When final approval is given to the customizations, the updates are moved to the Production Companies, keeping the QA and Production Companies in close synchronization.

Migration scripts and programs are tested in Migration Companies.

These relationships are illustrated in Figure 25.

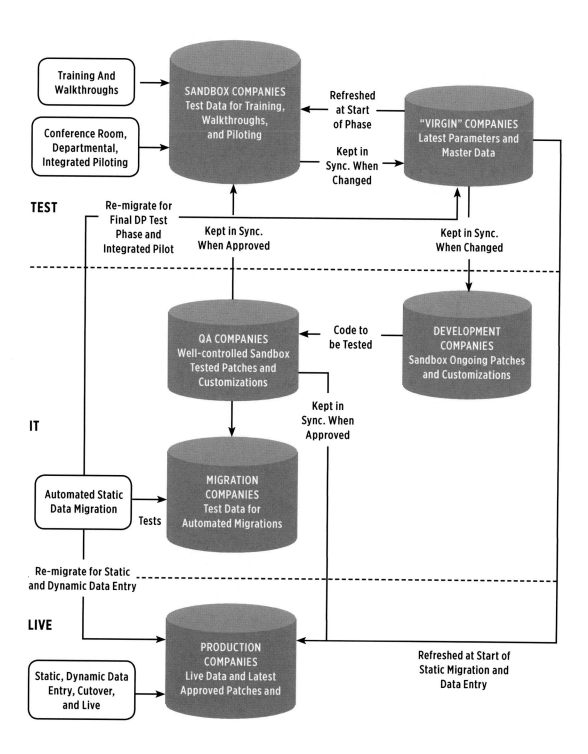

Figure 25. Migration Environment
Work within a simple framework of one or more companies: Virgin, Sandbox, Development, Quality Assurance (QA), Migration, and Production.

The Customizations, Interfaces, and Special Projects Phase
Customizations, interfaces, and special projects refer to those functions and enhancements not supported by the new ERP application. The development of these routines usually falls under the jurisdiction of the IT department.

The Timeline
Each of these endeavors may require a separate project plan. The core team training and Mapping Phases should be completed prior to starting any of the tasks in this phase.

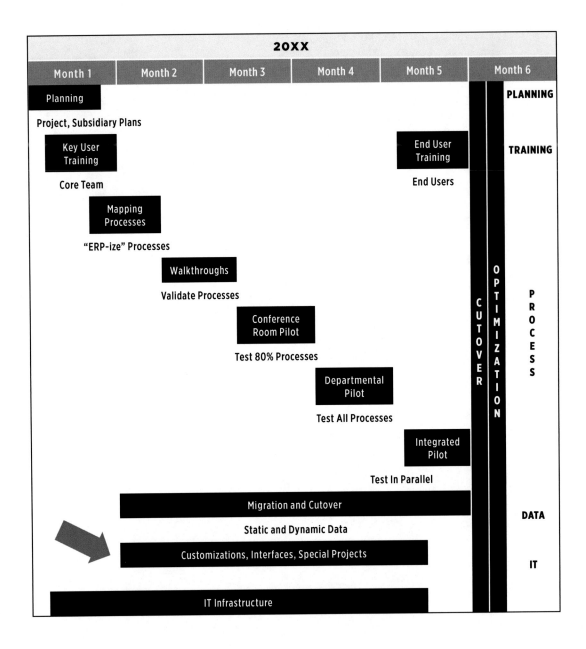

What This Phase Accomplishes

Customizations

Customizations are programming enhancements to the basic ERP application delivered by the software supplier.

In all implementations, the way of doing business and the ERP application modules must be adapted to work together. It is an unreasonable, over-simplification to assume that no changes will be made to the basic, out-of-the-box ERP software.

"Stand-alone" changes may be made without altering the original application source code or database structures. However, customizations which will change the code or the database may affect support from the software vendor and, as a rule, should be kept to a minimum. As the mapping process and walkthrough presentations proceed, the core team is responsible for identifying those changes to the basic software that make sense.

Interfaces

Depending on the number of legacy routines that will still be used, or the "bolt-on" programs and external business-partner electronic interfaces remaining after cutover, it may be necessary to interface with, import data to, and export data from, the ERP application database files. Each of these translation and interface programs require full IT sub-project definition and monitoring.

Special Projects

Like interfaces, special project needs (e.g., outside taxation software processes, non-integrated web-access, etc.) will require significant IT development and control.

The Tasks

1) Define specifications and monitor the development progress for software customizations, software interfaces, and other special projects

Complete these tasks for the beginning of the Integrated Pilot Phase.

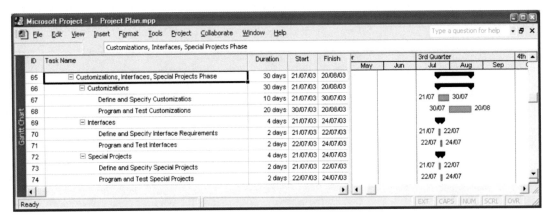

Figure 26. Customizations, Interfaces, Special Projects Tasks

The Deliverables

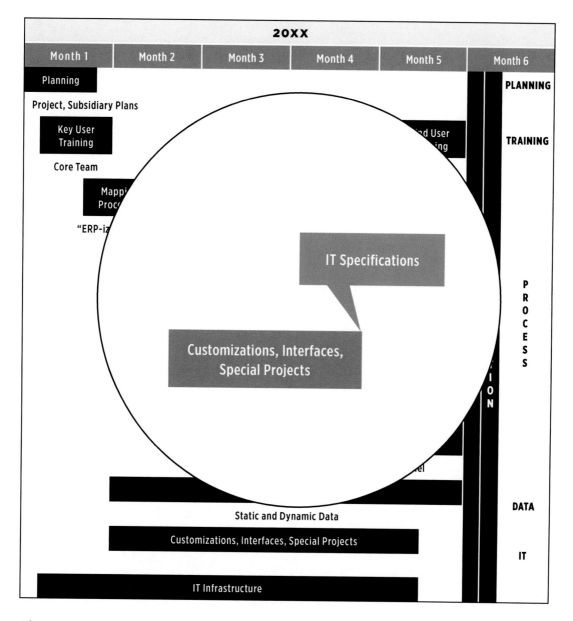

The Fourteenth Deliverable: IT Specifications

A set of specifications is needed for each and every request sent to IT for coding.

Formally documenting these requests serves many purposes:

1. IT developers need a much lower level of detailed specifications than those traditionally provided by the requester.

2. Documentation puts users and developers on the same page. If the requirement definition is not written, either party can inadvertently misinterpret what was said. By writing it down, changes in scope can be identified and controlled.

3. Both the requester and the developer are aware of the demand and can prioritize it.

4. The effort and associated cost of the request are known and allow management to schedule and track its progress.

Approval is also formalized and triggers the transfer of the program code from the Development Companies to the QA Companies, and from the QA Companies to the Production Companies.

A sample Customization request form follows.

IT Customization Request

Registration Information

Request Id:_____   Date Registered: _____

Functional Area: _____   Requester: _____

IT Coordinator: _____   Priority: _____

Request Description: _____

Development Process

Estimates: Detailed Specifications _____   worker-hours_____

             Programming_____   worker-hours_____

             Testing and documentation_____   worker-hours_____

Analyst 1: _____   Date Assigned: _____

Analyst 2:_____   Date Assigned:

Programmer 1:_____   Date Assigned: _____

Programmer 2: _____   Date Assigned: _____

Acceptance Process

Date Process Completed: _____

Analyst Signoff:_____

Date QA Completed: _____

Analyst Signoff:_____

Date User Accepted: _____

User Signoff: _____

Description/Objective of Work

_____
_____
_____
_____
_____
_____
_____
_____
_____
_____
_____
_____
_____
_____
_____
_____
_____
_____
_____
_____

Labeling

Session Identifier:_____

Menu location: _____

Session Form(s) Layouts

_____
_____
_____
_____
_____
_____
_____
_____
_____
_____
_____
_____
_____

Program Script Meta-code

_____
_____
_____
_____
_____
_____
_____
_____
_____
_____
_____
_____
_____
_____

Report Layouts

_____
_____
_____
_____
_____
_____
_____
_____
_____
_____
_____

Report Details

Breaks: _____

Sort Fields:_____

Totals and Sub-Totals: _____

Headers/Footers:_____

Report Script Meta-code

_____
_____
_____
_____
_____
_____
_____
_____
_____
_____
_____

The IT Infrastructure Phase

This phase relates to the IT infrastructures needed to support the new ERP application and its users. This phase is often split into separate series of sub-projects under the direction of the IT department.

The Timetable

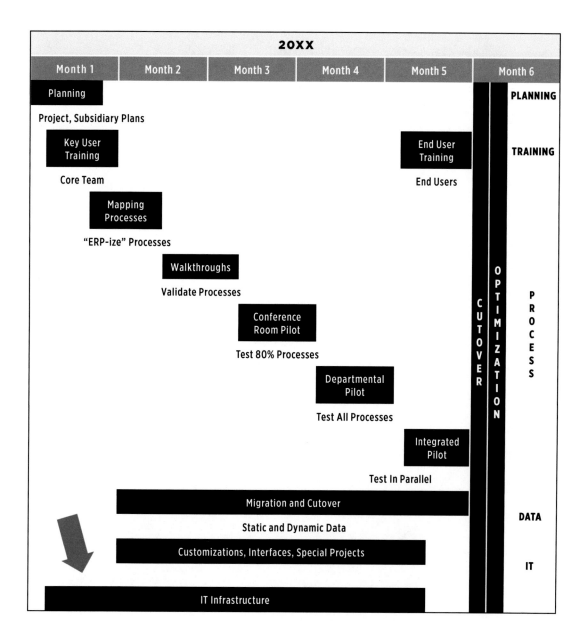

What This Phase Accomplishes

Sub-projects include, but are not limited to:

- The Network Infrastructure
  - LAN
  - WAN
- The ERP Application Server
- The ERP Database Server
- IT Change Management
  - Security control and maintenance
  - Patch and version control and maintenance
  - Workflow control and maintenance
- The ERP Application Production Infrastructure
  - Backup and recovery procedures
  - The printing systems
  - Batch job scheduling
  - "Bolt-on" software
  - TCPIP and Web interfaces
  - Stress testing
- The Help Desk
  - Support infrastructure
  - Training
- Disaster Recovery
  - Planning and documentation
  - Testing

Sample IT Infrastructure tasks are represented in Figure 27 through Figure 36.

The Tasks

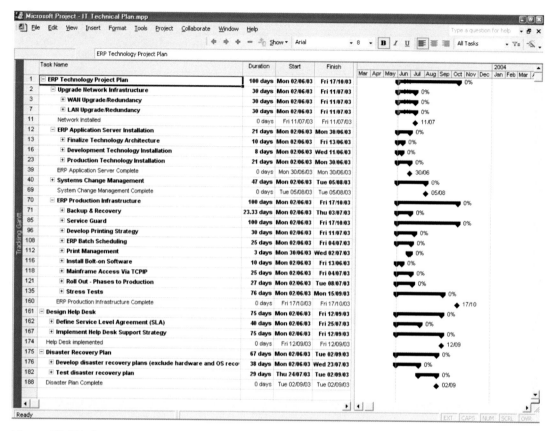

Figure 27. IT Infrastructure Tasks

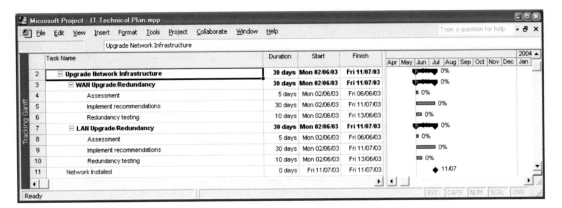

Figure 28. IT Network Infrastructure Tasks

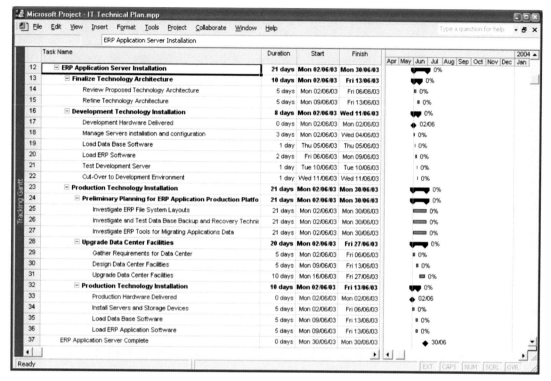

Figure 29. IT ERP Application Server Tasks

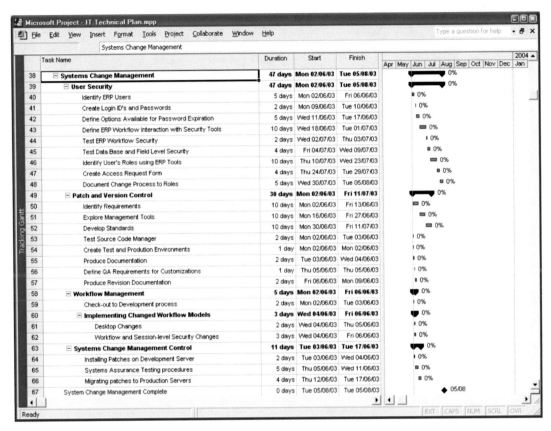

Figure 30. IT Systems Change Management Tasks

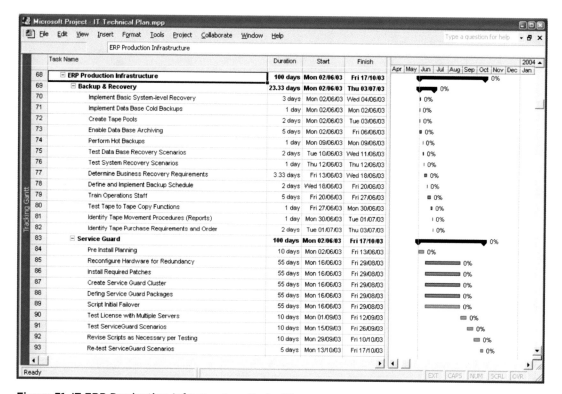

Figure 31. IT ERP Production Infrastructure Tasks (1)

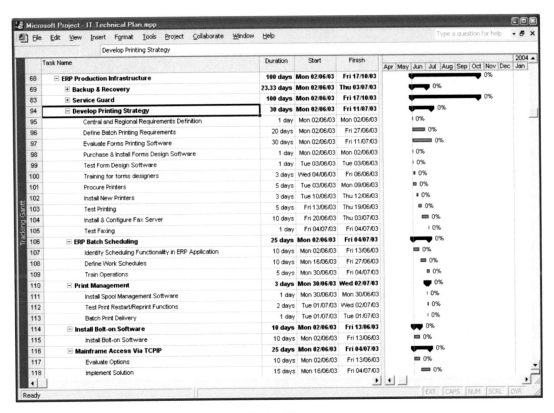

Figure 32. IT ERP Production Infrastructure Tasks (2)

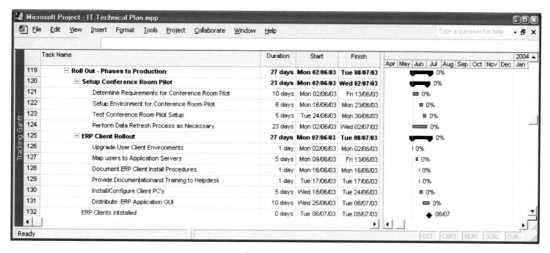

Figure 33. IT ERP Production Infrastructure Tasks (3)

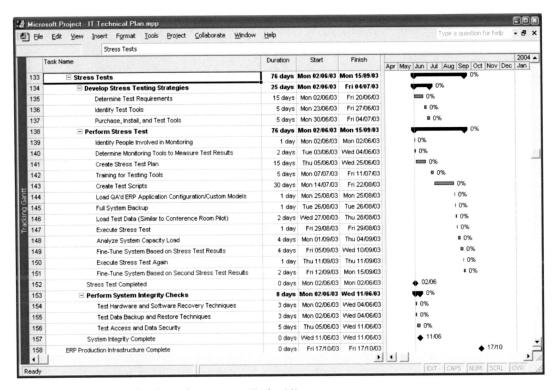

Figure 34. IT ERP Production Infrastructure Tasks (4)

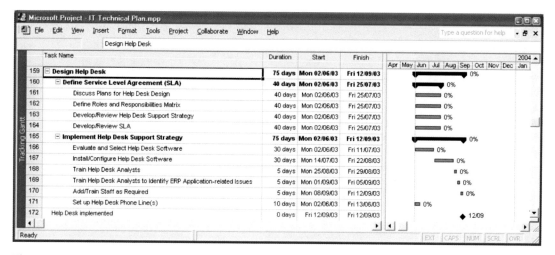

Figure 35. IT Help Desk Tasks

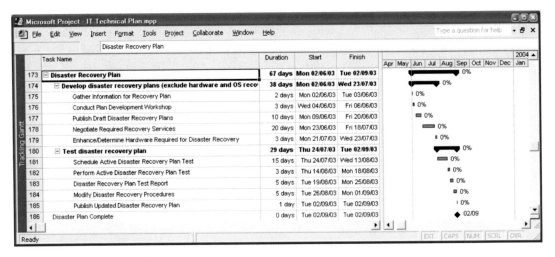

Figure 36. IT Disaster Tasks

The Deliverables

The Fourteenth Deliverable: IT Infrastructure Plans

The IT infrastructure plans and associated deliverables are extensive topics not covered in this book.

The Cutover Phase

There's actually a light at the end of the tunnel! All that's left to be sure the light's not a train heading our way is to get all the remaining data fully loaded into the production company, and balance it with the legacy system.

The cutover tasks referred to include:

- The long-term data entry and migration
- The short-term data entry and migration
- The cutover "weekend" tasks

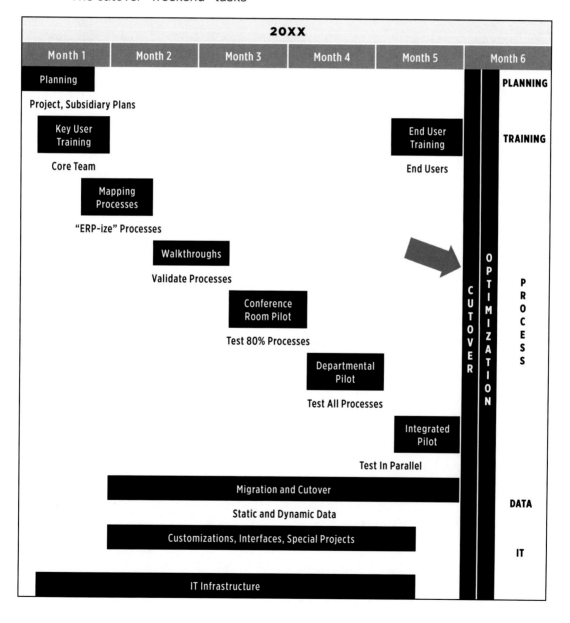

The Timetable

Although a fiscal year-end is thought to be ideal, cutover to the new ERP application is normally scheduled for a non-critical month-end weekend. The timetable depends on the volume of dynamic data to be transferred, the method of transfer, the requirements to conduct a physical inventory, etc.

Once again, for simplicity reasons, the examples assume a three-week, long-term dynamic data entry effort and a final, three-day cutover weekend.

What this Phase Accomplishes

Resources

At this point in the evolution of the project, the key users should possess a level of technical and business knowledge close to that of the consultants. Thus, if a core team presence is required in more than one geographic location, both key users and outside consultants may be used interchangeably.

The Cutover Phase includes three distinct events:

1.	Long-term dynamic data migration
2.	Short-term dynamic data migration
3.	Final tasks before going live

All dynamic data entry occurs in the production company files.

The Tasks

Pre-cutover events and tasks are as follows:

Cutover timetable

Using our six-month project example, the cutover countdown starts approximately three weeks prior to cutover.

Cutover Day minus 21 (CD – 21) through CD – 19

- By now, IT must have finalized job programming and testing the customizations, interfaces, and special projects; it must also have transferred these from the Development (or QA) Companies before Integrated Piloting (IP) commences. All special forms, special printers, specific security routines, and user desktops must be installed.

- The Sandbox Companies for IP must be refreshed from the Virgin Companies and, where appropriate, static data must be re-migrated

- The end user training continues

CD – 18 through CD – 15

- The IP begins in the Sandbox Company
-
- The end user training continues

CD – 14 through CD – 8

- Prepare the Production Companies by refreshing them from the Virgin Companies and, where appropriate, re-migrate the static data

- Print all legacy reports to obtain the pertinent data and documents for long-term dynamic data transfer into the new ERP Application

- Begin the long-term dynamic data entry into the Production Companies

- Continue the IP in the Sandbox Companies

- The end user training continues

CD – 7 through CD – 5

- Continue the long-term dynamic data entry into the Production Companies

- Continue the IP in the Sandbox Companies

- If needed, take a physical inventory in the legacy system

- End user training continues

CD – 4 through CD - 3

- Close as many open items as possible in the legacy system to prepare for the short-term dynamic data entry. This includes, but is not limited to:

    — Shipments to customers and to the shop
    — Receipts from suppliers and from the shop
    — Un-inspected receipts from suppliers and the shop

    — Open shop work orders
    — Unpaid supplier invoices
    — Customer cash receipts
    — Hourly labor transactions
    — Payroll transactions
    — Return incomplete warehouse "picks" to inventory

- Cease recording business transactions in the legacy systems in preparation for the month-end close

- Print legacy reports to obtain the pertinent data and documents needed to transfer the short-term dynamic data to the new ERP Application

- The end user training continues

CD – 2 through CD – 1

- Print the legacy month-end reports, adjust the balances where appropriate, and close all the legacy systems

- Transfer the legacy inventory balances and final costs into the new ERP application. Run the new ERP Application reports, and balance them with the legacy system

- Freeze these costs in new ERP Application

Cutover Day

- Begin the new business day using the ERP Application

- Physically locate the core team members next to the end users to supervise the departments' day-to-day business processes on the new ERP Application

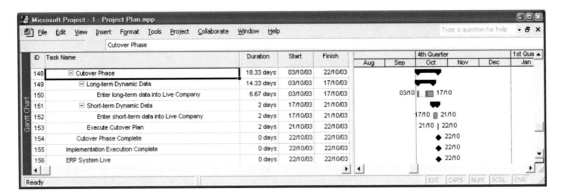

Figure 37. Cutover Tasks

The Deliverables

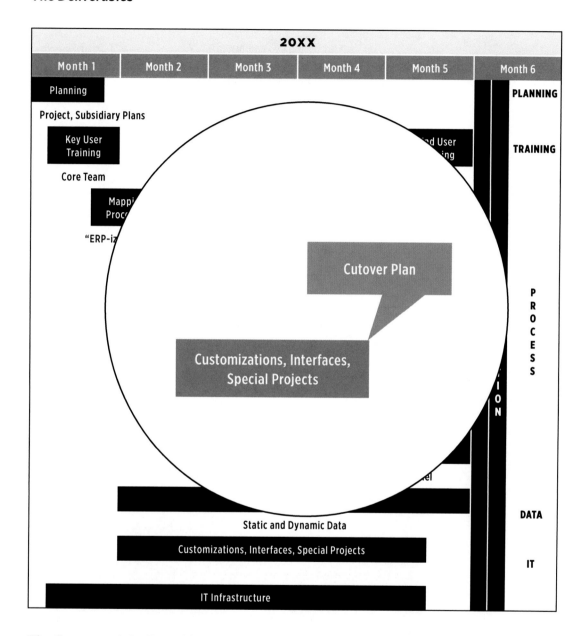

The Fourteenth Deliverable: The Cutover Plan

The cutover plan deliverables must include the following: a highly detailed list of the final tasks; resources for the core team assigned to these tasks; any limitations imposed by geographic constraints; a detailed schedule for the weeks leading up to the final weekend; and a day-by-day, hour-by-hour schedule for the final weekend.

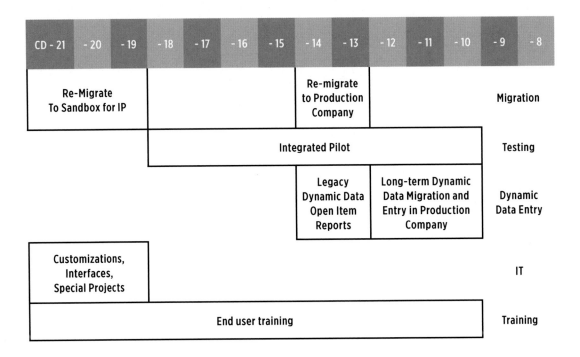

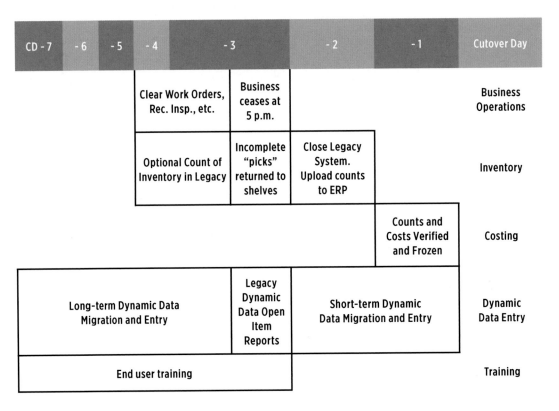

Figure 38. The Cutover Timetable

The Optimization Phase

Following cutover weekend and into the coming weeks, the end users wonder why they ever agreed to leave the comfort of their legacy systems as they tumble down the Emotional Curve towards "Desperation."

However, it will take several weeks to several months for things to settle down as end users take time to acclimatize themselves to new ways of working.

We now address the lower priority, outstanding gaps and issues that waited for cutover to complete.

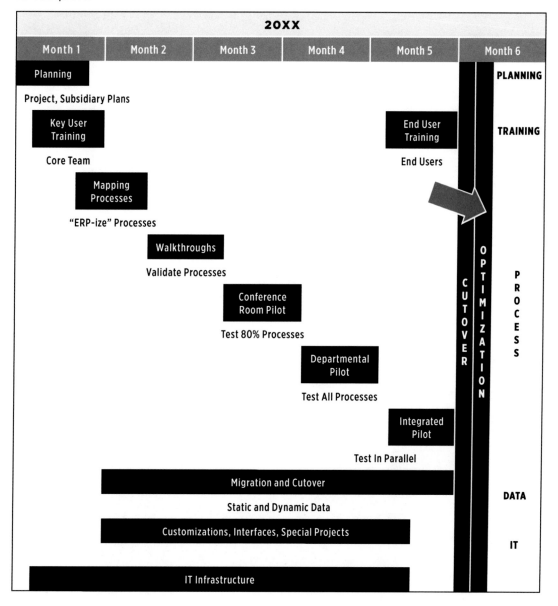

The Timetable
This optimization phase can last anywhere from one to several months.

What This Phase Accomplishes
In the first days after cutover, the core team must "baby-sit" the end users and coach them in their day-to-day routines. This will be a busy month for the core team due to last-minute, unforeseen problems, lower-priority issues left until after cutover, plus the first financial month-end close.

IT core team members are busy correcting glitches in custom code and are furiously writing reports and inquiries that slipped through the cracks during implementation.

Be prepared to suffer through several disquieting months until a semblance of normalcy returns to the business.

The Tasks

1. Complete tasks not required until first month-end close

2. Close first month-end

3. Address the remaining Priority 2 issues and gaps

4. Begin addressing Priority 3 issues, after-cutover system "nice-to-haves"

5. Update user documentation

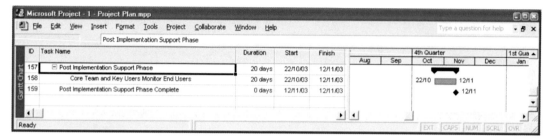

Figure 39. Optimization Tasks

The Deliverables

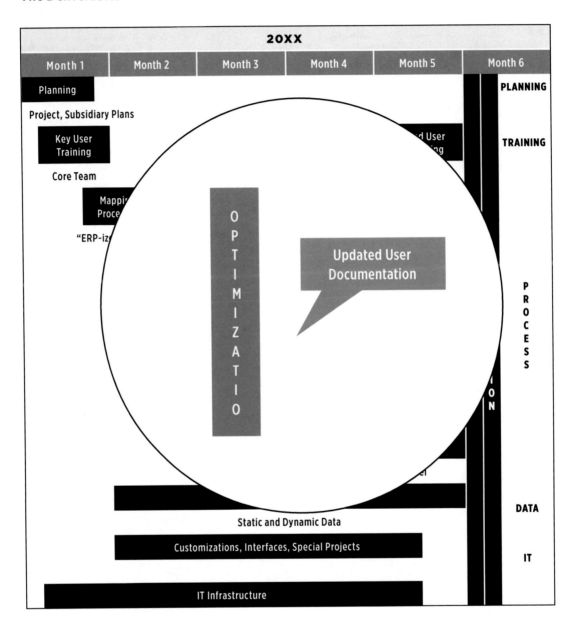

Updated User Documentation

This documentation must be kept up-to-date with ongoing changes to the business processes.

CHAPTER 4
SUMMARY

This section presents in one set of diagrams the phases, tasks, associated deliverables, as well as the resource responsibilities and target dates, based on the 6-month implementation model.

Figure 40 groups the Planning, Training, and Mapping Phases;

Figure 41 groups the Piloting Phases; and Figure 42 groups the End User Training, IT, Migration, Cutover, and Optimization Phases.

PLANNING, TRAINING, AND MAPPING

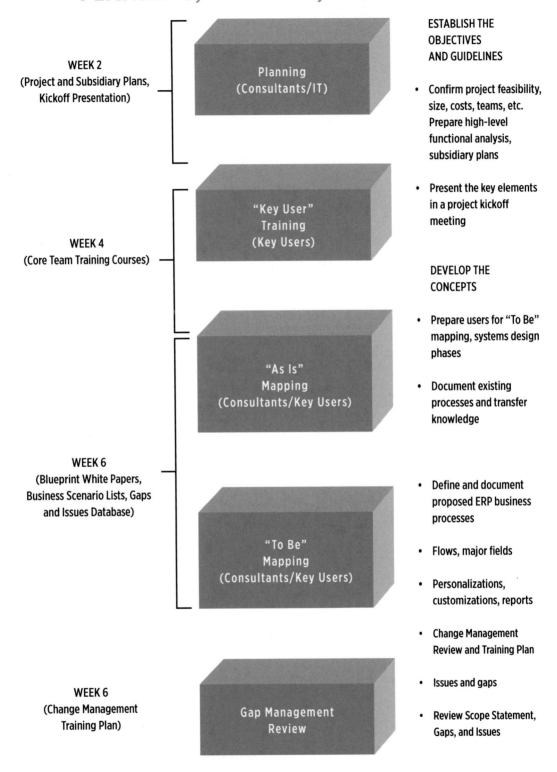

WEEK 2
(Project and Subsidiary Plans, Kickoff Presentation)

Planning
(Consultants/IT)

ESTABLISH THE OBJECTIVES AND GUIDELINES

- Confirm project feasibility, size, costs, teams, etc. Prepare high-level functional analysis, subsidiary plans

WEEK 4
(Core Team Training Courses)

"Key User"
Training
(Key Users)

- Present the key elements in a project kickoff meeting

DEVELOP THE CONCEPTS

- Prepare users for "To Be" mapping, systems design phases

"As Is"
Mapping
(Consultants/Key Users)

- Document existing processes and transfer knowledge

WEEK 6
(Blueprint White Papers, Business Scenario Lists, Gaps and Issues Database)

- Define and document proposed ERP business processes

"To Be"
Mapping
(Consultants/Key Users)

- Flows, major fields

- Personalizations, customizations, reports

- Change Management Review and Training Plan

WEEK 6
(Change Management Training Plan)

Gap Management
Review

- Issues and gaps

- Review Scope Statement, Gaps, and Issues

Figure 40. Planning and Mapping Tasks

PILOTING

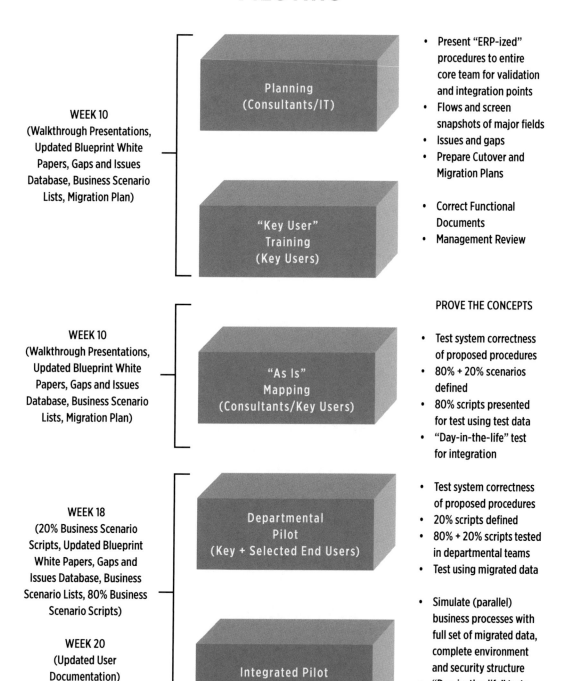

WEEK 10
(Walkthrough Presentations, Updated Blueprint White Papers, Gaps and Issues Database, Business Scenario Lists, Migration Plan)

Planning
(Consultants/IT)

- Present "ERP-ized" procedures to entire core team for validation and integration points
- Flows and screen snapshots of major fields
- Issues and gaps
- Prepare Cutover and Migration Plans

"Key User" Training (Key Users)

- Correct Functional Documents
- Management Review

WEEK 10
(Walkthrough Presentations, Updated Blueprint White Papers, Gaps and Issues Database, Business Scenario Lists, Migration Plan)

"As Is" Mapping (Consultants/Key Users)

PROVE THE CONCEPTS

- Test system correctness of proposed procedures
- 80% + 20% scenarios defined
- 80% scripts presented for test using test data
- "Day-in-the-life" test for integration

WEEK 18
(20% Business Scenario Scripts, Updated Blueprint White Papers, Gaps and Issues Database, Business Scenario Lists, 80% Business Scenario Scripts)

Departmental Pilot (Key + Selected End Users)

- Test system correctness of proposed procedures
- 20% scripts defined
- 80% + 20% scripts tested in departmental teams
- Test using migrated data

WEEK 20
(Updated User Documentation)

Integrated Pilot (End Users)

- Simulate (parallel) business processes with full set of migrated data, complete environment and security structure
- "Day-in-the-life" tests using selected actual legacy transactions

Figure 41. Piloting Tasks

END USER TRAINING, IT, MIGRATION, CUTOVER, AND OPTIMIZATION

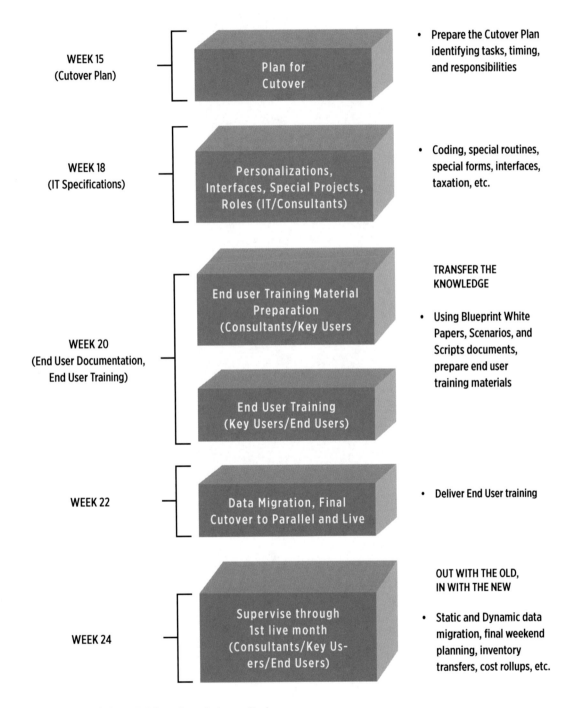

WEEK 15
(Cutover Plan)

Plan for Cutover

- Prepare the Cutover Plan identifying tasks, timing, and responsibilities

WEEK 18
(IT Specifications)

Personalizations, Interfaces, Special Projects, Roles (IT/Consultants)

- Coding, special routines, special forms, interfaces, taxation, etc.

WEEK 20
(End User Documentation, End User Training)

End user Training Material Preparation (Consultants/Key Users

End User Training (Key Users/End Users)

TRANSFER THE KNOWLEDGE

- Using Blueprint White Papers, Scenarios, and Scripts documents, prepare end user training materials

WEEK 22

Data Migration, Final Cutover to Parallel and Live

- Deliver End User training

WEEK 24

Supervise through 1st live month (Consultants/Key Users/End Users)

OUT WITH THE OLD, IN WITH THE NEW

- Static and Dynamic data migration, final weekend planning, inventory transfers, cost rollups, etc.

Figure 42. Training, IT, Migration, Cutover Tasks

CHAPTER 5
PROJECT MEETINGS

Meetings are necessary evils. On the one hand, they provide the mechanism to brainstorm as a group and to communicate the project status or issues of common interest across the functional teams. On the other hand, their misuse can drastically reduce productivity and adversely affect morale.

Take note. *A two-hour meeting attended by 20 people translates into one worker's entire week of indirect time charged to the project!* Make sure these meetings are worthwhile and productive.

Meeting Format and Minutes

Before calling a meeting

Be very clear about the outcome you hope to achieve and prepare a specific agenda, time frame, and list of the key stakeholders who should attend. Make sure to distribute copies of the agenda to all participants at the start of the meeting.

Set a time limit and encourage participants to mull ideas beforehand.

Start the meeting precisely on time and close the door so that latecomers are properly embarrassed. Sometimes we've required stragglers to pay a "fine," say a dollar per minute late, into a party fund.

During the meeting

Begin by reviewing the agenda, reminding participants of the goal, and requesting they stay on course. If someone veers off the topic, politely interrupt, restate the agenda, and ask participants to save their other topics for later.

If a secretary is not available to take minutes at meetings, ask for a volunteer or alternate the role among participants from meeting to meeting.

At the end of the meeting

Summarize the findings, come up with an action plan detailing next steps, and delegate these tasks to the participants. Set the date and time of the next meeting, if applicable.

Compile the minutes into a memo and send a copy to all interested parties. Ask for feedback or other thoughts that weren't discussed during the meeting.

Minutes

I've included a format for minutes that has proven successful over the years. As can be seen below, the complete history of an item is kept visible until the issue is resolved, at which time it is finally removed from the minutes.

ERP Core Team

Weekly Information Meeting (No. 12)

MINUTES

February 12, 20XX

Present: Peter G., Ralf S., Paul L., Frank C., Claudio P., Luc D., Hubert F., Mario V., François P., Lisa M.

Absent: Joyce G., Alain C., Kevin S.

Copy: Malcolm M.

Item	Due Date	Item Description	Resp.
General			
12.1		The coming tasks are:	
		• complete existing flows (acct., service, inventory)	ALL
		• prepare business scenarios with Joyce G. and enter test data for pilots	PG, FP, MV
		• department refreshers on system and pilots while we redefine flows in the group	ALL
		• integrated pilot	ALL
Distribution and Inventory			
7.2	20XX/01/06	Frank C. will present the HO Distribution flows	FC
8	20XX/02/06	Frank C.'s Distribution presentation was first class. Items to be discussed or amended include:	
		• freight handling	
		• partial shipments	
		• drop shipments to regions	
		• reallocation of goods to more urgent MR's	

Refers to the meeting number during which the minute was initially created

Refers to the meeting number during which the minute was updated

Item	Due Date	Item Description	Resp.

Project and Installation

Item	Due Date	Item Description	Resp.
6.1		Mario V. will present the installation process at the next meeting	
7		The installation overview was given by Mario V. and the following was noted: • installation orders for training should be created and tracked through the process • to enforce the discipline to follow the procedures will require convincing upper management of the financial benefit.	

Manufacturing

Item	Due Date	Item Description	Resp.
11.2		Kevin S. presented the manufacturing cycle. Questions arose as to the long-term direction this department will be taking vis-à-vis subcontracting, etc.	
	20XX/02/13	The flowcharts will be transferred to computer by the next meeting	KS
		The following was noted: • manufacturing is complicated and re-engineering will be difficult but the system features will greatly benefit the company • the returns to sub-contractors and international shipping procedures will be detailed further	KS

The next meeting will take place on Wednesday, Feb. 19, 20XX at 10:00 in the main conference room.

Kickoff Meeting

A carefully constructed, well-supported kickoff meeting enhances the project's chances of success. Although it does not necessarily ensure success, an inadequate one will almost certainly hinder the project.

The meeting's duration will vary according to the size of the project, but usually merits a half-day. For wider-ranging ERP projects, nearly all employees are involved, and careful preparation is critical.

A Formal Meeting

Ideally, make it a formal, well-organized meeting that is held away from the hustle and bustle of the workplace. This demonstrates management's commitment to, and the significance of, the project. After all, any payback activity important enough to involve key people and managers for an extended period of time should merit an appropriate launch.

The Agenda

A good kickoff meeting is somewhat similar to a coach's pre-game locker room spiel. It should also contain some pep-rally elements! Most ERP situations require a half-day, comprised of two segments.

1. **The General Meeting**

Attendance is mandatory for most senior management, the steering committee members, and the project core team. Preferably, if possible, all employees should attend. A proposed agenda is:

   a. An opening statement by the CEO (or other senior manager) underscoring senior management's commitment of time, availability, and finances to the project
   b. The presentation of the Project Charter with the expected results and benefits
   c. An explanation of the roles of:
      i. The executive steering committee
      ii. The core team
   d. A short introduction of the project team members

      e. A presentation of the project elements, notably:
          iii. the ROI
          iv. the Milestone Deliverables methodology
          v. the core team commitment
          vi. the "Emotional Curve"
          vii. the related project timeline
      f.   The inherent risks and risk mitigation strategies

2.   The Core Team Kickoff Meeting

Another session for the core team builds a foundation for their working relationship. Topics include:

      a. An expansion of the mission statement into the measurable success factors (MSF) and associated strategic business accomplishments (SBA) developed in the project plan
      b. A project name, if one hasn't already been decided
      c. A detailed presentation of the methodology and its related project timeline
      d. The responsibilities, roles, and duties for core team members
      e. How issues will be recorded and tracked

The kickoff meetings are crucial to set the expectations and tone of the project. It's not just "show time." Senior management must communicate their long-term commitment and full support during the anticipated highs and lows of the project.

Steering Committee Meeting

The purpose of the steering committee meetings is:

1. To bring the committee members up-to-date on the general project status
2. To discuss scope or financial changes that need stakeholder approval
3. To escalate issues to the committee that cannot be resolved by the core team

The steering committee meeting normally takes no more than half a day. Traditionally, it is held monthly. Of course, it can be assembled any time there are major issues that require immediate attention.

All steering committee members are required to attend. From time to time, outside managers are invited when issues require their presence.

A typical steering committee meeting agenda is:

1. Opening comments and addition of other items to agenda (CIO)
2. Prior business - review of minutes (CIO)
3. Project status report (Project Manager)
   i. General
   ii. Timeline and resource review
   iii. Financial review
4. New business (Project Manager)
   iv. Scope changes
   v. Issue escalation
5. Other items (CIO)
6. Closing comments and next meeting date (CIO)

Core Team Meeting

The purpose of the core team meetings is:

1) To bring the core team up-to-date on the general project status
2) To discuss issues of common concern that cross functional boundaries
3) To discuss issues that cannot be resolved alone by the functional teams
4) To identify issues that must be escalated to the steering committee

This meeting is held weekly.

Attendees are the core team members. From time to time, end users and managers are invited when issues require their presence.

A typical agenda is:

1. Opening comments and addition of other items to agenda (Project Manager)
2. Prior business - review of minutes (Project Manager)
3. Project status report (Project Manager)
   i. General
   ii. Timeline and resource review
   iii. Financial review
4. New business (Key Users)
   iv. Cross-functional issues
   v. Unresolved issues
   vi. Issue escalation
5. Other items (Project Manager)
6. Closing comments and next meeting date (Project Manager)

Functional Team Meeting

The purpose of the functional team meetings is:

1) To bring the functional team up-to-date on the status of the functional area's deliverables and task status
2) To discuss open issues
3) To identify issues that must be escalated to the core team
4) To identify issues that must be escalated to the steering committee

This meeting is held weekly.

Attendees are the functional team members. From time to time, end users and managers are invited when issues require their presence.

A typical agenda is:

1. Opening comments and addition of other items to agenda (Key User)
2. Prior business - review of minutes (Key User)
3. Status report for the functional area (Key User)
   i.   General
   ii. Timeline and resource review
4. New business (Key User, Selected End Users, Outside Consultants)
   iii. Open issues
   iv. Unresolved issues for escalation
5. Other items (Key User)
6. Closing comments and next meeting (Key User)

Cutover Meeting

During the final weeks before cutover, the project needs to be increasingly micro-managed. This is not because we trust our key users any less, but because the slack on pre-cutover tasks is negligible and any slippage introduces unrecoverable delays.

At this point in the project life cycle, the integrated pilots are just finishing. The core team and end users are diligently following their training plan while others are entering static data into the production companies. The IT team is putting the final touches on the customizations, tweaking the hardware settings, and tightening user security. The activity is intense and there is precious little time for distractions.

However, based on the six-month template project we've established, twice-weekly cutover meetings should start at approximately Cutover Day minus 30 days (CD – 30). At CD – 15, these should be increased to every other day, and to daily from CD – 7 through CD itself.

The core team members are the cutover meeting attendees. From time to time, end users and managers are invited when issues require their presence.

These meetings must be kept short and to the point. Emphasis must be placed on the key schedule dates.

The typical cutover meeting agenda is:

1. Prior business - review of minutes (Project Manager)
2. Status report for the functional area (Key User)
3. New business (Key Users, Selected End Users, Outside Consultants)
   i.   Open issues
   ii. Unresolved issues needing escalation
4. Other items (Project Manager)
5. Closing comments and next meeting (Project Manager)

CHAPTER 6:

MISCELLANEOUS TOPICS

The War Room

Formal team meetings, ad-hoc meetings, formal and informal training sessions require a proper venue. A multi-purpose War Room should be made available for the full duration of the ERP implementation.

At a minimum, it should have seating capacity for the entire core team.

This room requires access to the system network for presentation software, the ERP application software, the Internet, teleconferencing, etc. Each training participant requires a PC or terminal; printers must be available nearby.

There must also be access to an overhead projector with a PC feed, a white board and flip charts.

Rapid ERP Implementations

Remember the three dimensions of project success mentioned earlier: completing all project deliverables 1) on time, 2) within budget and, 3) achieving a level of quality that is acceptable to sponsors and stakeholders.

A rapid ERP implementation is one where the timetable to complete the project is shrunk significantly from the ideal plan. The budget will be affected in a positive way, but the risk of delivering an inferior product is greatly increased.

Since we can assume that the ideal plan's schedule has little room for paring down, on which phases can shortcuts be taken without drastically affecting the quality of the end product?

Before answering that question, we want to emphasize that there are phases and associated deliverables where shortcuts should never be taken:

1. End user training
2. End user documentation
3. Data integrity – migration and data entry

This is simple common sense. Poorly trained users, poor documentation, or bad data can never be worked around, and a deficiency in any of these areas will ensure a project's failure.

A pre-requisite to the compression techniques described below is the correct choice of key users for a Rapid ERP Implementation.

On a Rapid ERP Implementation, key users must be extremely computer-literate, ERP-knowledgeable, cross-functional, business-process savvy individuals who also possess an ability to think outside the box.

That's quite a "mouthful." Accordingly, these types tend to be completely indispensable to running the business, so it's a challenge to free up their time for the project in the first place. That's the bad news for senior management. The good news is that the project that the company needs so badly will be up and running that much faster.

Each of these key users should be able to manage more than one functional area at a time, but they will have to spend *110%* of their time on the project. This will shorten the training, mapping, walkthrough, and piloting phases considerably.

These are the phases that can be compressed:

Less Emphasis on Introductory Training
Because the Core Team is made up of ERP-knowledgeable and computer-savvy individuals, the overview and navigation training modules in the Core Team Training Phase can be shortened or even eliminated.

Reduce "As Is" Mapping Time
The "as is" Mapping Phase serves several purposes: 1) to document existing processes as a reference point for the "to be" maps; 2) to familiarize core team members with details of departments outside their day-to-day operating scope; and 3) to provide outside

consultants with an understanding of the current business environment.

These key users are savvy about cross-functional business processes. Rough "as is" flows will suffice. Also, their scenario lists will be sufficient to make sure no business processes are left uncovered. Although it's still important to provide consultants with a detailed understanding of the current environment, this can be gained through informal interviews, tours, and discussions with these key users.

Streamline the Walkthrough Process
The walkthroughs are the forum where the conceptualized business processes are presented for the core team's review. With fewer cross-functional key users, points of contention and inter-departmental integration disconnects are reduced significantly. The resulting presentation time for the walkthroughs is shortened.

Shorten the Departmental Pilots
These key users have a detailed understanding of their business processes. This results in quicker development and testing of the 80% and 20% scenario scripts. If they alone assume this responsibility, this will eliminate the requirement to train other selected end users to help complete this task.

Work-Arounds
When managing highly stretched goals, the project can afford no slippage along the critical path. In Rapid Implementations, many of the gaps and issues are resolved with "work-arounds," some incurring a higher business-processing overhead than costly and time consuming, long-term fixes. These solutions are delayed until after cutover.

Micro-Manage to Maintain Focus
With such an emphasis on meeting aggressive dates and targets, almost all tasks lie along the critical path and can afford no slippage. It falls to the project manager to maintain the team's focus on the dates and deliverables. Since the key users are too deep in the forest to see the trees, this over-attention to detail forces the project manager to micro-manage more of the tasks than normal.

Extend the Optimization Phase
Be prepared to spend time fine-tuning the ERP system after cutover! The work-arounds must be replaced with proper functionality and correct business processes. The reports previously placed on the back burner now need to be developed.

Hints for Outside Consultants

Let's change our perspective to that of the outside consultants hired on the project as core team members. For me, this is a return to a comfortable viewpoint.

As mentioned earlier, at a strict minimum, the outside consultants must possess the following attributes:

1. General knowledge of business and business operations
2. Leadership qualities and a flair for facilitating discussions
3. Excellent communication skills
4. Significant product-related functional and technical knowledge

General Business Knowledge

At the outset, the consultants' understanding of the client's corporate goals and business operations will be superficial at best. However, their depth of general business knowledge plus their prior experience on similar mandates should help them quickly absorb new information and accelerate their learning curve.

During the first few weeks of becoming familiar with the new environment, the consultants should be good listeners and ask a lot of questions. They must avoid trying to impress the client too much, too early, and must not spout off how things were done differently or better elsewhere.

Leaders and Facilitators

With their in-depth knowledge of the product and the methodology, consultants are in a strong position to direct people - training key users, facilitating mapping sessions, and orchestrating implementation phases. They are also in a position to correct people - revising business processes, and advising system usage. This is not easy! Without direct authority over the core team members, the consultants' communication and diplomatic skills must be honed to perfection.

Communicators

From the top to the shop, the consultants must establish friendly, professional, relationships with the company's management, the core team, and the end users. Many times, I have been asked by the CEO how things are going or how his or her *own* project manager is doing! Be ready to respond honestly and fairly. By the same token, I am privy to lots of office politics and chatter and must avoid commenting or getting too close for my own good.

As subject matter experts, consultants sometimes disagree with their clients. Be tactful, as this is a delicate balancing act. When the choice is one of disagreeing with the client to correct an error, or taking the wrong path to avoid making waves, the professional must make the former choice every time. It's always good business sense to put the project's success first and foremost.

Product Knowledge

The consultants must know their product from the inside out. Each and every one of them must be prepared to roll up his or her sleeves and get down and dirty in the trenches with the rest of the core team and end users. This must be a true *team* effort!

When they arrive on-site, the consulting staff must be integrated immediately into the team as peers. It is the role of the project manager to *delineate* the responsibilities and tasks of the consultants to avoid possible confusion and conflict with the rest of the team.

Finally, as consultants, we must not *do* everything ourselves. It is the key users and end users themselves who must put together the project deliverables even if it would take less time for the consultants to prepare them on their behalf. I've seen too many projects where the IT team and the consultants end up doing all the work. The result is a system nobody wants that's forced down the throats of the users.

The Consulting Project Manager

Before all else, the consultant project manager must be a master communicator and diplomat.

The toughest challenge he or she will face will be to offer guidance and assistance to the steering committee, the core team, and the end users without projecting an image of one-upmanship over the client's project manager or appearing to undermine his or her authority.

Establish the reporting hierarchy with the client immediately upon arrival. Make it clear that, as outside project manager, he or she reports to, and takes direction from, the client project manager. And make sure he or she respects that relationship during the life of the project.

LIST OF FIGURES

INDEX

ABOUT THE AUTHOR

Peter Gross, B.Sc., M.Eng.

Peter – a leading implementation expert according to the High Court of New Zealand – has built a worldwide reputation as an authority on delivering successful ERP implementation projects.

Peter's two greatest gifts are his ability to break down the complex into small targets, and his ability to get people to collaboratively work towards those targets.

Peter is Pemeco Consulting's President and Founder. He has more than 35 years' experience leading complex enterprise software, company integration, and process reengineering projects. Peter has personally delivered more than 100 major implementation projects, including the successful rescue of several failed projects initially managed by global consulting firms. In addition to his project management expertise, Peter is an authority on manufacturing business processes, including: MRP II, master production scheduling, capacity requirements planning, logistics, costing, and supply chain management.

Made in the USA
Middletown, DE
20 December 2020